Snakes, Sunrises, and Shakespeare

Snakes, Sunrises, and Shakespeare

How Evolution Shapes Our Loves and Fears

To Richard, Highly valued colleague

Gordon H. Orians

Gordon H. Orians

THE UNIVERSITY OF CHICAGO PRESS | CHICAGO AND LONDON

The University of Chicago Press, Chicago 60637
The University of Chicago Press, Ltd., London
© 2014 by The University of Chicago
All rights reserved. Published 2014.
Paperback edition 2015
Printed in the United States of America

23 22 21 20 19 18 17 16 15   2 3 4 5 6

ISBN-13: 978-0-226-00323-8 (cloth)
ISBN-13: 978-0-226-27182-8 (paper)
ISBN-13: 978-0-226-00337-5 (e-book)
DOI: 10.7208/chicago/9780226003375.001.0001

Library of Congress Cataloging-in-Publication Data

Orians, Gordon H., author.
 Snakes, sunrises, and Shakespeare : how evolution
shapes our loves and fears / Gordon H. Orians.
     pages cm
 Includes bibliographical references and index.
 ISBN 978-0-226-00323-8 (cloth : alkaline paper) —
ISBN 978-0-226-00337-5 (e-book)
1. Emotions. 2. Evolution (Biology)—Psychological
aspects. I. Title.
 BF531.O73 2014
 152.4—dc23
                    2013019829

♾ This paper meets the requirements of ANSI/NISO
Z39.48-1992 (Permanence of Paper).

Contents

1

Whistling

for Honey

How old is the human sweet tooth, and why do we crave sweet things?

As it happens, these two questions will take us far from our distant ancestors and their hunger for wild honey. They will lead us to still other questions, questions that have to do with the emotions the natural world arouses in us—yearning and revulsion, joy and fear—and how those emotions have shaped every aspect of our lives.

But our story starts here, following a prehistoric hunter somewhere in Africa, whistling for honey.

Until humans figured out how to refine sugar from plants such as sugarcane and sugar beets, we had to steal from creatures skilled at concentrating the nectar of flowers into a rich source of food: honeybees. Our ancestors learned how to rob the nests of wild bees at least twenty thousand years ago—rock art of that age survives in Zimbabwe (figure 1.1) The image left behind on the rock face clearly shows a person smoking a hive to get honey.

Our African ancestors must have found honey a sweet temptation: it was nutritious, delicious, and easy to digest. But bee colonies are uncommon on the African savanna. In order to find and exploit them, prehistoric humans relied on an unusual partner that also benefited by leading them to the hives. That partnership has persisted into modern times; we can still witness it today in several African tribes, among them the Boran of northeast Kenya.

Just before setting off in search of honey, Boran honey hunters give a specific, loud whistle, known as the *fuulido* (figure 1.2). If they're in luck, there will be a return call from a bird with a *fuulido* of its own. The caller is a greater honeyguide, a small bird; its Latin name, *Indicator indicator*, reflects its value to humans. The honeyguide repeats its distinctive

1

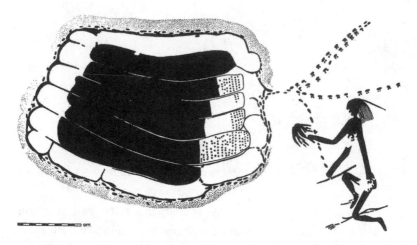

Figure 1.1. Rock painting showing a honey hunter using smoke on a wild hive, Toghwana Dam, Zimbabwe.

"follow me" call, and begins to escort the Boran hunters to a bee's nest, pausing frequently to allow the hunters to catch up. Most astonishing, upon arriving at the tree where the bees have nested, the honeyguide perches and sings a special "indication" song. The bird escort remains nearby as the hunters disperse the bees and claim the prize of the honeycomb. In gratitude, they always reward their navigator by leaving behind some honeycomb. Unlike most birds, honeyguides can digest wax; they feast on it along with the honey and bee larvae lodged in the comb,[1] but honeyguides are too small and weak to open bees nests. They depend on humans to do it for them, just as people depend on honeyguides to help them find the nests.

This unusual mutually beneficial partnership features in the myths of many African savanna-dwelling tribes. Our ancestors probably gorged on honey whenever they could find it. Honey was a nutritional and energy bonanza, a precious fuel for our large hominid brains. But the rarity of hives in the savannas made it impossible to gather enough to grow fat on this rich resource.

By contrast, walk the aisles of a modern supermarket and count the staggering variety of sweet and sugary foods. We modern humans are blessed and cursed by uninterrupted access to sweets, and we have become slaves to our sweet tooth. Hardwired with our ancestors' cravings for sugar-rich foods, we're unable to resist. We are also adapted to

Figure 1.2a. A Hadza hunter named Darabe whistling for a honeyguide.

Figure 1.2b. Hadza boy eating a nutritionally rich honeycomb.

an environment where food was sometimes abundant and sometimes scarce. When it is abundant, we lay on fat for the future hard times. Today, hard times rarely come. As a result, obesity is now a serious health problem throughout the developed world and an increasing problem in developing nations.

It turns out that a fondness for sugar is just one trait our ancestors bequeathed to us. Our ancestors' responses to environmental challenges—unpredictable sources of food, ever-present predators, extremes of weather—have molded our modern emotional lives. They are a central theme of this book. Evolutionary psychologists tell us that whenever we're incited to act by strong emotions, positive or negative, chances are good those actions were of great evolutionary importance. Responding appropriately to stimuli meant the difference between surviving or not, leaving offspring or not.

Our ancestors came to prefer or "like" beneficial objects and events in nature that increased their chance of surviving and passing on their genes to their children—in short, evolution by natural selection. Conversely, they came to avoid or "dislike" objects and events that were threatening and decreased their chances of surviving and reproducing. Over time, these likes and dislikes became wired in the human brain. As a result, we have a taste for honey and a nearly universal fear of carnivores with big teeth. Science allows us to trace these ancient emotions and find the adaptation in what we find beautiful and what fills us with revulsion and fear. We can better understand how we interact emotionally with our environment by viewing our behavior through an evolutionary lens that focuses on our ancestors.

This book records the results of my efforts to find out how our emotional lives bear the imprint of decisions our ancestors made long ago on the African savanna as they selected places to live, sought food and safety, and socialized in small hunter-gatherer groups. I hope to convince you that the impressions are many and deep and the rewards for this new understanding as useful as the honeyguides are to the Boran people.

How I Came to These Questions

My search for an environmental basis to emotions and aesthetics began the year I was seven, when I discovered the world of birds. My family rented a cabin on a lake in northern Wisconsin, and I was captivated by the call of the common loon. I soon began to record observations of birds I saw; those notebooks still sit on the shelves of my university office. When I was about thirteen, I joined birders in the Milwaukee, Wisconsin chapter of the Audubon Society. A few of them were professional ornithologists. At some moment, I put two and two together—people were actually paid to study birds! I decided then that I would go to college, major in biology, and become a professional biologist. I did exactly that. I became a behavioral ecologist because I was interested in the decisions birds like honeyguides must make to be successful—how they select habitats, search for food and decide what to eat, chose their mates, and invest in their offspring.[2]

In my young adulthood I never questioned my strong attraction to birds. I simply enjoyed them. But as I matured as an evolutionary biologist, I began to think deeply about human emotional responses to nature. My thoughts, as well as those of many other people, were stimulated by Edward O. Wilson's *Sociobiology: The New Synthesis,* published in 1975.[3] Wilson helped me recognize that the decisions I was studying with birds govern our lives as well. My curiosity has led me in surprising directions and caused me to learn about topics that I knew little or nothing about, but all examined through an evolutionary lens.

Like most other people, I have been deeply moved by the roar of ocean waves, lightning-filled skies, and claps of thunder. Views of sunsets and beautiful spreading trees have given me pleasure. The smell and sight of rotting flesh repulse me. Yet, *why* we respond emotionally the way we do to the cheerful songs of birds or the glow of sunshine—this was a foreign topic, one that my scientific colleagues and I seldom pursued in depth.

Emotions through an Evolutionary Lens

Evolutionary biologists like me came to the party late. For centuries, artists and philosophers had explained the human emotional response to beauty by pointing to human culture and works of art. We must look primarily at our responses to human creations, they argued, to understand our feelings and the basic structure of our emotional lives. Daring to explain our emotional responses to nature in evolutionary terms can provoke outrage among artists and nature lovers. In the nineteenth century, John Keats despaired that Isaac Newton had forever sullied the beauty and mystique of rainbows by explaining how they were formed. In his poem *Lamina* (1820) he complained that Newton had destroyed a rainbow by unweaving it. Keats was not alone, then or now (figure 1.3). Even today, most people resist having their emotions explained because they believe that explaining them will destroy their inherent wonder. Such defiance is so pervasive that Richard Dawkins wrote *Unweaving the Rainbow: Science, Delusion, and the Appetite for Wonder* (its title taken from the line in Keats's poem), which addressed this false perception.[4]

Scientists marvel at rainbows even though they know that physical processes generate them, just as most of us intensely enjoy physical intimacy even though we understand its biological function. Yet, many people are hostile to the idea that our emotions evolved because they positively influenced the survival and reproductive success of our

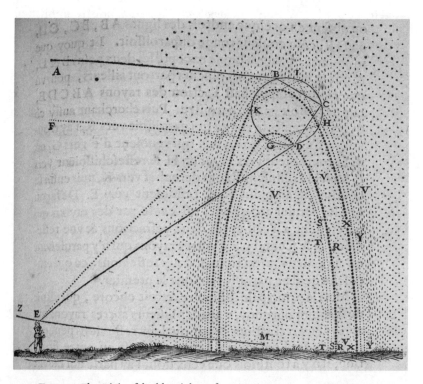

Figure 1.3. The origin of double rainbows from René Descartes, *Les Météores*, 1637.

ancestors. They resist the idea that science might better explain our emotions than philosophy, art, or common sense.

But what are emotions? We all know, or think we know, much about emotions. We know that they powerfully affect our behavior and thoughts; that some of our emotions are pleasant, others are not. Yet, as Fehr and Russell describe, "everyone knows what an emotion is, until asked to give a definition."[5] We struggle to define them. Fortunately, we can set aside the definitional problem and study emotions directly. In this book I will draw upon self-reports (rating scales, questionnaires) and upon physiological measures of emotions. By examining this evidence we will develop a deeper understanding of what emotions are than we could gain by trying to come up with a definition.

To understand our emotional responses to the environment we shall explore in greater depth why we even have emotions at all. That exploration, covered in the next two chapters, will set the stage for later discussions of how humans perceive the environment with our senses and why our emotional responses to nature are so diverse. I hope to demonstrate that an evolutionary approach to human behavior gives us a creative way to think about our emotional lives and a way to answer many questions we have about them. The evolutionary approach to why we have the emotions and aesthetic responses we do differs strikingly from the prevailing view held by Western academics. We will be concerned primarily with the so-called basic emotions—pleasure, anger, fear, pain, surprise, and disgust. We will pay little attention to the social emotions—love, guilt, shame, embarrassment, pride, envy, and jealousy.

In this book I share with you the results of my personal odyssey to understand our emotional responses to nature and how scientists explain them. Our emotional roots lie in the African savannas, back where we first followed honeyguides to a sweet feast of honey, while keeping an eye out for lions that might dine on us. We will examine how an understanding of our evolutionary history helps explain why we respond emotionally, why the reasons for our responses are often hidden from us, and how our brains evolved to make decisions that improve our survival and reproductive success. Understanding these emotions is particularly important today when we live and respond emotionally to a complex environment that differs dramatically from the one our ancestors lived in until very recently. I will also explore how the experience of our emotional responses influences both what we do to the environment as well as what the environment does to us.

The current traits of organisms are the result of their history because evolutionary processes cannot anticipate what the next week, century, or geological era will look like. For modern humans, this means that some of the emotional responses to environments that served our ancestors well may no longer be adaptive in today's industrial societies. For example, we still dream about primeval spiders and snakes rather than about more pervasive modern threats like guns, nuclear weapons, and climate change. Evolutionary insights may help us identify and explain why some elements of our responses are no longer adaptive.

My explorations of human-environment interactions have been both challenging and enjoyable. I hope to convey at least some of that pleasure and excitement to you and to demonstrate that we may be able to use our improved understanding of the roots of our emotional lives to make better decisions today. And understanding ourselves better is, itself, a rich source of pleasure.[6]

2

Ghosts of the African Savanna

We respond with strong feelings to the world around us. We react to some objects, places, and events with pleasure, and call them beautiful; others arouse feelings of fear, disgust, or horror, and we call them ugly. But why do we have this aesthetic sense?

Intellectual thinkers have speculated about the origins and meanings of an aesthetic sense at least since the early sixth century BC in Greece, but Baumgarten's *Aesthetica*, published in 1750, established the science of the sense experience.[1] It was Baumgarten who gave us the word "taste" in the sense of the human ability to judge what is good. This view of an aesthetic sense opened the way to the scientific study of emotions in the nineteenth century by such men as Charles Darwin, William James, and Wilhelm Wundt.[2]

As early as 1785 the Scottish philosopher Thomas Reid recognized that our emotions might have evolved because they were useful:[3]

> By a careful examination of the objects which Nature hath given this amiable quality (of beauty), we may perhaps discover some real excellence in the object, or at least some valuable purpose that is served by the effects it produces upon us. This instinctive sense of beauty, in different species of animals, may differ as much as the external sense of taste, and in each species be adapted to its manner of life.

In this remarkable passage, written seventy years before Charles Darwin published *On the Origin of Species*, Reid proposes that the aesthetic senses of animals are likely to be related to their relationships with their environment. He suggests that animal and human emotional responses evolved because of benefits that accompanied them.

A generation later, Darwin would show how natural selection, acting over millions of years, could produce structures beautifully fitted for a specific task: eyes for seeing, ears for hearing, wings for flying. But could the human mind, with its powerful emotions, evolve in the same way? Darwin thought it had. Thirteen years after the publication of his 1859 masterwork, he published *The Expression of the Emotions in Man and Animals*. It was lavishly illustrated, with engravings showing the human facial muscles, a dog with its hackles raised, a grimacing baboon, a sulky chimpanzee—as well as many images of people. Some of the images were photographs by a French physiologist, Guillaume-Benjamin-Amand Duchenne, published with his 1862 book *Mécanisme de la Physionomie Humaine, ou Analyse Electro-physiologique de l'Expression des Passions*.

Duchenne used a device he had originally developed to investigate the muscles that control the hand to find the muscles responsible for creating particular expressions. He applied an electrical current to the facial muscles of a number of different test subjects. He also took photographs of the same people with blank expressions, and others where they were attempting to simulate expressions without the aid of the probes (figure 2.1).

At this time, Darwin was forming his own ideas about the expression of emotions in humans. He was fascinated by Duchenne's photos and wondered whether the expressions Duchenne had elicited from his patients were universal, whether certain muscle movements of the face always accompanied the same emotional state, in other words, whether there were universal expressions of human emotions. Did a certain grimace always mean disgust? To find out, he conducted experiments on his own—on the guests at his own dinner parties.

> It fortunately occurred to me to show several of the best plates, without a word of explanation, to above twenty educated persons of various ages and both sexes, asking them, in each case, by what emotion or feeling the old man was supposed to be agitated; and I recorded their answers in the words which they used. (*Expression*, 14)

The first visitors Darwin tested were his cousins, who came to dinner on March 22, 1868. A week later in London he hosted another party, and got the opinions of a number of his naturalist friends. He found that his guests accurately inferred the emotions of the people in the photos. Part of what we know about the parties has come down to us from witnesses.

Figure 2.1. Some of Guillaume-Benjamin-Amand Duchenne's images of facial expressions from Darwin's *Expression of the Emotions in Man and Other Animals*, 1872.

Darwin conducted one of his experiments during the visit of the Harvard botanist Asa Gray. His wife, Jane Gray, wrote a letter to her sister describing how funny the dinnertime experiment was and how the guests went off afterward and practiced making faces in front of mirrors.

Darwin was among a group of nineteenth-century thinkers focusing on the age-old problem of emotions. Philosophers across the ancient world had struggled with the question of what emotions are. But with the dawn of modern psychology and the advent of the machine age, scientists had an increasing array of techniques and technologies at their disposal. They asked study subjects to complete adjective checklists, rating scales, and questionnaires, or simply asked them to describe what they felt. They could record and quantify emotions using instruments to measure heart rate, respiration, skin conductance, muscle tension, and blood pressure.[4]

But alone among these investigators, Darwin was looking at human emotion through an evolutionary lens. Viewed from that perspective, emotions were not the gift of a divine creator but a legacy of our animal origins. Even as natural selection had shaped our ancestors' brains and bodies, so had evolution shaped their emotions. If we wanted to see how emotions might have evolved, Darwin argued, we could form hypotheses by looking at ourselves and at other animals: our pet cats and dogs, or the chimpanzee in the local zoo.

> No doubt as long as man and all other animals are viewed as independent creations, an effectual stop is put to our natural desire to investigate as far as possible the causes of Expression. . . . With mankind some expressions, such as the bristling of the hair under the influence of extreme terror, or the uncovering of the teeth under that of furious rage, can hardly be understood, except on the belief that man once existed in a much lower and animal-like condition. . . . He who admits on general grounds that the structure and habits of all animals have been gradually evolved, will look at the whole subject of Expression in a new and interesting light. (12)

As he states in this quote, Darwin concluded that our emotions and how we express them have deep evolutionary roots. Indeed, he judged this to be so obvious that in the final paragraph of his book he states, "but as far as my judgment serves, such confirmation was hardly needed." Today, no serious scientist contests that conclusion.

Emotions' Evolutionary Roots

Darwin's perspective suggests that we should look for ways that particular emotions and the actions based on them could have helped our ancestors survive and reproduce.[5] For example, it is obvious that those ancestors who enjoyed sexual intimacy would have passed more copies of the genes influencing those preferences to subsequent generations than those who did not enjoy, and, hence were less stimulated to seek out, sexual partners. By the same logic, individuals who were attracted to and hence settled in safe environments rich in resources (caves, water, game) should have left more offspring than individuals who were attracted to and settled in inferior habitats. Those offspring would have inherited copies of the genes influencing those preferences.

For these reasons, an evolutionary perspective on aesthetics suggests that beauty and ugliness are not intrinsic properties of objects. Rather beauty and ugliness arose from interactions between traits of objects and the human nervous system. In this view, beautiful objects were ones that, if we responded to them positively, improved our lives—increasing our likelihood of surviving, winning a mate, and leaving offspring.[6] Ugly objects were ones that interfered with or impeded some aspect of living. In other words, we should view our responses to beauty and ugliness by asking, what do they accomplish? We will make more progress in understanding our emotions and our aesthetic responses if we ask, "how might these responses have helped our ancestors solve problems?

From the Prairie to the Playground

A May morning in 2010 found me standing alone on windswept prairie in eastern Montana, observing a herd of pronghorn antelope. I approached as quietly as I could, but not quietly enough. Suddenly one animal sensed my presence, and in an instant they had all bounded away. In a few seconds the herd had covered a vast distance and were almost lost to sight. The pronghorn, *Antilocapra americana*, is North America's fastest land mammal, able to sustain speeds of fifty-nine to sixty-five miles per hour for many minutes. They are so fast, in fact, that no predator around—not bears, coyotes, or wolves—can catch them (figure 2.2). Watching them, a curious evolutionary biologist might ask, if they can

Figure 2.2. Pronghorn antelope (*Antilocapra americana*) running in a grassland, Yellowstone National Park, Montana.

outpace every predator, why do pronghorn continue to race across the prairie, expending huge amounts of energy? Or, as an inquisitive child might put it, if nothing can catch them, why do pronghorn run so fast?

The answer is the key to understanding an important theme of this book: we can understand the behavior of contemporary animals only by recreating in our minds their past environments and the forces that shaped their responses. The remarkable speed of the American pronghorn is probably a legacy of selection by American cheetahs (*Miracinonyx* spp.) (figure 2.3) that became extinct during the last ice age.[7] The swift cheetahs the pronghorn coevolved with are long gone, but the pronghorns are still speedy. If I were to try to explain their incredible speed and endurance without understanding the past environment, my explanation would doubtless be wrong.

We can apply this same logic to human responses to the environment. Until as recently as one hundred thousand years ago, humans lived in small groups on the African savanna. They continued to live in small groups and sought similar environments as they dispersed out of Africa. Human nature has been molded by the responses of our ancestors to events in that long-ago environment. It follows that we will better understand who we are, our emotional response to our current environment, indeed our most basic emotions, if we first grasp what our ancestors did, why they did it, and the consequences of their actions.

For example, consider children playing on a playground at recess. Watch and notice where the children head—which kids start a game of dodge ball or tag, which head to the swings, and which immediately start to climb. Which kids are the first onto the climbing structure? Would you be surprised to know it is the girls?

We now call them play structures or climbing frames, but in an earlier generation playground climbing equipment was known as the jungle gym or the monkey bars, usually made of metal piping from which children could dangle by their knees or swing, hand over hand, ape or monkey fashion. That name, the monkey bars, gives us a clue to the puzzle of why girls climb more than boys. It turns out to be an environmental

Figure 2.3. The extinct North American cheetah (*Miracinonyx*) is shown hunting a pronghorn antelope (*Antilocapra americana*) in the Great Basin region during the late Pleistocene.

ghost from our past, a legacy of our primate ancestors. We are apes, not monkeys, so the bars should have been called "ape bars," but you get the point.

We came down from the trees, adopted a bipedal lifestyle, and took advantage of all it offered. But hominid females never lost the ability to climb trees. They were lighter and more agile than the males, but also smaller and more vulnerable, more at risk from savanna predators. Trees offered the only refuge from large carnivores. Females would have climbed not just to escape predators, but also to collect fruit and nuts and take advantage of the shade from the hot equatorial sun. Thus tree climbing was adaptive for females, less so for males. Males eventually became less adept at climbing, this argument goes, but climbing by smaller and more vulnerable females was favored much longer.

What evidence do we have that this story is true? First, girls *do* climb more than boys. Girls climbed more frequently than boys, regularly much more often, on elementary school playgrounds, and they fell less often. It appears that they are better adapted for climbing, compared to boys. Women have a wider range of foot motion than men, allowing them to grip tree branches with all four limbs. Male foot morphology indicates that hominid males climbed less often.[8]

There is another line of evidence. Most savanna primates sleep in trees; so did our ancestors. If females were more likely to sleep in trees than males, they should be particularly alert to attacks by predators from below. Males, on the other hand, sleeping on the ground, should be more attentive to side attacks. Three- to four-year-old boys and girls are equally fearful at night, but boys are more afraid of danger to one side—the monster in the closet—while girls are more fearful of something dangerous below—the monster under the bed.[9]

When we observe girls at recess swinging from the monkey bars with ease, are we witnessing a ghost from our environmental past? I believe so, but I have come to this view of human emotion gradually. I was raised the American Midwest, in a Protestant family. I was taught to seek meaning by appeals to spiritual rather than evolutionary forces. Only when I matured as a biologist did my thinking shift to an evolutionary perspective. I discovered why ghosts of environments past, like the speed of pronghorns on the prairie, might reside in my own psyche and shape my construct of reality. Novel as it may seem, delving into past human life on African savannas is fundamental to an inquiry into our current emotions and the way we live our twenty-first-century lives. So let's journey back and take a closer look at the lives of our hominid ancestors about fifty thousand years ago.

Surviving on the Savanna

The social environment we live in today has existed for a remarkably brief period of time. Until thirty-five thousand years ago, we lived in small, hunter-gatherer groups. The next milestone in our evolution, the domestication of cereal crops and livestock, occurred only about ten thousand years ago. The Industrial Revolution started within the last two centuries. We have been truly modern for fewer than ten generations. So the behavioral responses that enabled our hunter-gatherer ancestor to survive on the plains of Africa served us well until very recently.

To survive on the African savannas, let alone breed and stay healthy, our ancestors needed many skills. During their daily forays in search of food, they needed to navigate through the landscape, recognize objects, understand and make tools like nets and spears, judge distance, avoid mammalian predators and venomous snakes, identify edible plants, and capture animals. When eating, they needed to avoid disease-causing organisms and consume a balanced diet. With experience, they needed to decide which foraging efforts repaid their energy expenditure and which did not. They needed to select mates of high reproductive value and successfully court them, while avoiding incest. Successful males had to prevent their mates from conceiving by a rival male. Parents needed to attend to alarm cries, detect when their children needed help, and be motivated to offer that help. Living in groups meant they had to interpret social situations correctly, recognize faces and emotions, help relatives, deter aggression, maintain friendships, cooperate, and make effective trade-offs among many of these activities.[10]

Some of these behaviors, such as detecting when our children need help and interpreting social situations correctly, are still vital. Others, such as identifying wild food plants and avoiding venomous snakes and dangerous mammals, no longer hold much value in contemporary urban society. Still other behaviors, such as the ability to maintain a good nutritional balance, are declining in our modern society even though we still need them.

We can divide these many skills of hunter-gatherer life into five categories: (1) shelter, finding a place to live; (2) safety, avoiding injury or death by protecting themselves from bad weather, dangerous animals, or people; (3) nourishment, acquiring sufficient quantity and quality of food; (4) friends, choosing good associates; and (5) achieving contentment, a goal that clearly helped support the other four categories.[11]

Hunter-Gatherer Society

By 1.6 million years ago, our ancestors could walk upright rapidly and efficiently. This posture reduced heat stress and water loss, so our ancestors could travel long distances during the day when dangerous carnivores were resting. They could seek and find food in places too dangerous to visit at dawn, at dusk, or at night. So the hunter-gatherer lifestyle was born. These small bands of close relatives were seminomadic, moving seasonally to follow their food sources. Few places on Earth provided enough resources to enable people to settle in one place year-round. Members of the band helped one another survive danger, sharing food, child-rearing duties, knowledge, and skills. Men hunted and fought males of other groups, while women gathered edible plants, trapped, fished, and harvested eggs (figure 2.4).

Walking upright also freed our ancestors' hands for throwing things (such as nets and spears) and processing food (butchering animals, cleaning fish, cracking nuts). With their improved hunting skills, these early ancestors added animal protein and fat to a plant-dominated diet.[12]

As I stated in chapter 1, the currency of evolution is reproductive success—the genetic contributions of individuals to later generations. To contribute genes to its descendants, an organism must survive to reproduce; some of its offspring must do the same. Our helpless infants and slowly maturing children require intimate care and prolonged attention from parents, especially the mother. Until very recently human infants depended on their mothers' milk about three years on average.[13] Infants remained in almost constant physical contact with adults during

Figure 2.4. BaAka women eating foraged fruit on a hunting trip, Central African Republic.

the first several years of their lives. Their mothers or other adults carried them while they searched for food.

Hunter-gatherer groups were small and women bore a child only once every four years or more. Therefore, a group would have had only a few children. They would have socialized primarily with children of other ages: they would have learned much through contact with older children. As they grew they would have helped find food and care for infants. The same-aged peer groups in which the youths of today socialize have existed only a short time.[14]

The hunter-gather diet was probably reasonably well balanced but food was often scarce, delaying the onset of puberty. Once women began to ovulate, they were likely either pregnant or lactating most of the time. Males may have associated with particular females long enough to attempt to monopolize their reproductive careers, sometimes for many years. Males interacted with infants and children much less than women did, but men provided food and protection. Some of our ancestors appear to have practiced infanticide, perhaps to allocate scarce resources or eliminate deformed or sickly infants.[15] Intertribal conflict appears to have been common, flaring up into warfare.

Soon after our ancestors began to walk upright their brains enlarged rapidly. A larger brain can assimilate, interpret, and respond to more information. Primates that live in large social groups have larger brains relative to their body size than ones that live in smaller social groups. Living in a complex social group appears to favor larger brains.

But a larger brain comes at a cost. For example, selection for giving birth to a large-brained infant conflicted with selection for a narrow pelvis to support bipedal locomotion. The shape of the female pelvis changed to yield a larger opening; the shape of heads of babies evolved to make it easier for them to pass through the birth canal; but the changes were insufficient for human babies to be born as developmentally mature as infant gorillas and chimpanzees. As a result of this trade-off in the design of the female pelvis, we are born with very immature brains. The type of brain growth that is completed during pregnancy in gorillas and chimpanzees continues for a year after birth in human infants. Human children have an unusually long period of dependency during which they learn about their social and natural environments while still sheltered and cared for by adults. To determine what legacies from our ancestors reside in our psyches, it would help to know what they paid special attention and responded to.

Ghosts of Environments Past

The biological world is replete with "ghosts." There are ghosts of habitats past, ghosts of predators past, ghosts of parasites past, ghosts of competitors past, ghosts of fellow humans past, and ghosts of meteors, volcanic eruptions, hurricanes, and droughts past. How can we identify ghosts and determine why they survive in our minds? We can narrow our search for environmental ghosts by recognizing that environments bombard us with much more information than an individual can assimilate and use. Information overload is not new. It has plagued our ancestors for millions of years. Fortunately, we can ignore much incoming information because it has little value. The evolutionary responses to information overload are neural programs that emphasize or deemphasize some of the arriving information. These neural programs function much like filters, screening out irrelevant information and letting through only the important data. By having such filters we exhibit what is called *biologically prepared learning*.[16] That is, our minds filter out extraneous information, paying attention to and storing information that influences decisions affecting survival and reproduction.

Behaviors may persist because not enough time has elapsed for genetic mutations and genetic drift to erode them or because the genes may also code for a trait that is adaptive and actively selected for. Recent investigations show that behavioral patterns can persist long after natural selection no longer favors them. Arctic species of moths that have long lived in bat-free environments still have anti-bat defensive behaviors.[17] Anti-snake behavior has persisted in North American ground squirrels for at least several hundred thousand years in areas lacking snakes. Only Arctic ground squirrels that have lived for three to five million years in snake-free environments fail to respond to them. A ground squirrel generation is one year; three hundred thousand years scaled up from ground squirrel generations to human generations would be five million years.

No terrestrial snake in Madagascar can inflict a life-threatening bite. Snakes have not seriously injured anyone on the island during the two thousand years that humans have lived there. Yet most Malagasy display the nearly universal fear of snakes. A highly feared snake that represents the serpent in the Garden of Eden features in early Malagasy translations of the Bible.[18] (We'll learn more about the long human relationship with snakes in chapter 5.)

The typical objects of fears and phobias pose little danger in modern societies, yet our fear and avoidance of the feared object persists.[19] On the other hand, we have developed only weak fears to dangerous objects that only recently have become environmental threats. Behaviors clearly have persisted in animals for thousands of generations after they had little or no effect on fitness. Nonetheless, some scientists doubt that Pleistocene ghosts reside in the human mind. Jared Diamond has argued, "Humans spread out of Africa's savannas at least 1 million years ago. We have had plenty of time since then—thousands of generations—to replace any original innate responses to savanna with innate responses to the new habitats encountered."[20] However, we cannot use general plausibility arguments to tell if environmental ghosts exist. To demonstrate or reject Pleistocene ghosts we must test specific hypotheses experimentally.

Since leaving Africa and adapting to hot and cold climates and new diets, we have evolved different body sizes and shapes, skin pigmentation, and immune responses. We evolved the ability to digest some of the plants found in new geographical regions, allowing us to adapt to and exploit new sources of food.[21] Nevertheless, ghosts can persist for many generations if they aren't selected against or have no influence on fitness. In subsequent chapters we will explore the possible existence of ghosts by assessing whether the environmental clues to which our ancestors responded are similar enough today that formerly adaptive responses still serve us today. We will find out that some of them are but that some are not. Specific hypotheses will help explain why they persist.

3

The High Cost of Learning

Why should we have ghosts of the savanna in our heads? In part it's because learning is expensive, and not just in the sense of college tuition. From birth, humans spend massive amounts of time gaining, interpreting, and using new information. As a species, we are vain about our vaunted capacity to learn, but learning is very costly. Our central nervous system is only 2 percent of our adult body weight, but it accounts for some 20 percent of our metabolic requirements. It is biologically "expensive" to build and maintain a system that can acquire knowledge through the senses and store them in memory, retrieving knowledge as needed to make informed decisions. For many years we spend most of our waking hours learning things. Many of us spend decades of our lives in school. If we could dispense with learning, we could use that time to do something else. This is why evolutionary biologist George C. Williams asserted that all elements of behavior that can be instinctive—stored genetically—are likely to evolve to be instinctive.[1]

Learning isn't just costly and time-consuming; it can be risky. The price of not learning life's lessons is often high. We may fail to learn something important or learn the wrong thing. We have all heard dark jokes about people "taking themselves out of the gene pool." But what if knowledge were available to us from birth, like preinstalled software? It turns out that some knowledge is. We introduced that knowledge in chapter 2 where we called it ghosts of past environments. Ghosts of past environments are important from the moment we are born. Before we can stand, walk, or talk, our infant brains possess a kind of genetic memory, innate knowledge. Innate knowledge is that body of things we know about our world without having direct experience of them.

For example, we seem intuitively to understand concepts of space, time, and causality that we could not deduce from personal experiences. Prussian philosopher Immanuel Kant, in his monumental *Foundation of the Metaphysics of Morals*, concluded that we could not comprehend the environment unless we entered the world already understanding these relationships.

Recent research shows that Kant was correct. For example, the brains of rat pups moving through a maze for the first time light up in areas where the brains of experienced older rats keep information about direction.[2] That is, the brains of naive rat pups are primed to attend to and understand spatial relationships the first time they view them. Human infants also have an innate sense of spatial relations.[3] We also display innate knowledge by the disgust we experience when we look at pus and rotted, odorous flesh. We did not know that the microorganisms that cause these symptoms existed until very recently, but we innately respond as if we do know what causes pus and rot.

Our minds and those of other animals possess innate knowledge because having that kind of knowledge enabled our ancestors to respond appropriately to new situations. Evolutionary biology allows us to see innate knowledge as a legacy from our deep evolutionary past, but where does that knowledge come from?

The Origins of Innate Knowledge

An animal having a nervous system with internal models that embody "expectations" about the world can judge the significance of new information and use it to respond appropriately. That is, it can behave as if it anticipates the results of its responses the first time it makes them. Natural selection has been the "judge" of how good those responses have been. None of this requires (nor does it preclude) that an individual be cognitively aware of the consequences of its decisions. William T. Powers explored this perspective on knowledge in his classic book *Behavior: The Control of Perception*.[4] Identifying and explaining what kinds of innate knowledge we have will be a major focus of this book.

Recognizing that biological evolution is the source of innate knowledge gives us three important insights. First, we can understand why our knowledge fits our environment, why we seem to be able to understand the world. Second, an evolutionary perspective provides an answer to the puzzle of how we know so much despite having such limited personal experiences. Third, evolution explains why our knowledge reflects the environments in which our ancestors evolved. Our minds are adapted to the external world before we emerge from the womb, even as the hoof of a horse is adapted to the prairie ground before the horse is born and the fin of a fish is adapted for swimming before the fish hatches.

Behaviors as Adaptations

Like hooves and fins, behaviors may be adaptations that are based on innate information. When members of a species confront an environmental problem or opportunity for sufficient generations, the response that evolves can result in genetic changes. In this book I sometimes use a verbal shortcut by saying that natural selection has favored a "behavior," but natural selection cannot do that. It can select only for mechanisms that produce behavior. Behavior does not differ in this regard from other features of organisms, such as hearts, wings, hooves, fins, and digestive systems. For all these traits, natural selection can favor only molecular mechanisms that produce some outcome, whether that is a behavior or a wing, but the accurate statement is awkward and long. The shortcut is handy, as long as we remember what it really means.

Behavioral adaptations often are specific; selecting a habitat, finding food, and courting, to mention only a few, present unique challenges. A single general-purpose behavior is unlikely to be an efficient means to solve every challenge.[5] So a suite of behaviors evolves, rather than a single behavior to suit all circumstances.

We do not need to know how an adaptive behavior arose to study it. We simply need to identify why natural selection favored the mechanisms that produce it. We need not demonstrate that it has a genetic basis. In fact, we know little about the genetic basis of most adaptations. Today we know which genes govern the development of the fetal heart, but from its structure and operation we knew that the human heart was adapted for pumping blood long before the rediscovery of Gregor Mendel's work on peas in 1900 ushered in the beginning of the Age of the

Gene. We do not doubt that a bird's wings are adapted for flying, even though we have much to learn about the genetic programs that build them (figure 3.1).

Ethical considerations limit the experiments we can perform on people, but we can study human adaptations in many other ways. Fossils leave traces of behavior (figure 3.2). We are the only living hominid species, but we can gain insights by studying our nearest relatives, the great apes and other primates. Human cultures are real-world "experiments" in how to deal with the many different environments in which we live. Comparing the behaviors that differ among cultures (for example, courtship rituals) with those that are relatively universal (disposal of the dead) can help us discover adaptations. Studies of hunter-gatherer cultures help us understand how and why our ancestors made the decisions that governed their lives. We can also use the changes in our behavior as we age, become increasingly mobile, and encounter the environment in new and different ways, to test hypotheses about adaptations.[6]

In the chapters that follow I focus on the adaptations that motivate and guide our responses to the physical, biological, and social environments in which we live and through which we move. As we will see, that set of adaptations is very large.

Figure 3.1. A murmuration of starlings signal winter is on its way as they arrive at Gretna in Scottish borders.

Figure 3.2. Pleistocene human prints, western New South Wales, Australia.
Mary Pappin (*foreground*) found the human prints while working with Bond
University professor Steve Webb. They are cleaning up the site with children
from the three traditional tribal groups of Willandra Lakes. The site promises
to be the world's largest collection of Pleistocene human footprints.

Seeing through Hominid Eyes

Every time we look at an object or scene, we see at it through the eyes of
our hominid ancestors, unconsciously assessing what we could do with
it or in it. We may also evaluate the consequences of those actions. We
ask of an object not only, *what is it?* but also, *what's in it for me?* How easy
would it be to enter that environment, explore it, and find my way back?
What would I gain? Is that tree a source of fruit, a refuge from predators,
or a lookout post? A river might be a source of water, a channel to move
along, or an obstacle to be crossed.[7] Flocks of birds or herds of mammals
might signal an environment rich in game. Without our being aware
of it, our nervous system is busy filtering incoming data, suppressing
or enhancing information, checking the accuracy of perceptions, and
evaluating information. Because our modern lifestyle has arisen in a
mere blink of evolutionary time, we all do this with brains adapted for
life on the African plains.

The benefits things offer are captured by the concept of *affordance*, introduced in 1979 by the psychologist J. J. Gibson.[8] Affordance is what an object or environment offers to an individual at a particular time. We cannot simply tally the attributes of an environment and quantify its affordance because affordance varies with season, weather, and an individual's current needs.[9]

The evolutionary outcome of making and acting upon those evaluations is that we intuitively recognize and seek out high-quality environments. That is, we have evolved to seek out environments and objects that benefit us and ignore or avoid ones that don't. The evolutionary good news is that many of the things we enjoy doing are likely to be good to do.

Much of the environmental data upon which our hominid ancestors depended concerned the location of objects in space. Survival depended on knowing those locations: Where were prey animals yesterday? Where did I cache the food I could not carry back to camp? Where are the trees with ripe fruit? Where are safe hiding places that I may need to use in an emergency? Locating resources is important, but for a highly social species, so is understanding and assessing the social environment. We assess both kinds of environments using both innate and new information. Let's explore how that works.

Adaptive Responses to Nature

Genes store knowledge about ancestral worlds; they prepare individuals for circumstances they have never faced before. Genes also influence what things we most easily learn about and remember.[10] Many parts of our natural environment remain much as they were throughout human evolution. If we were magically transported to an African savanna at the time when the brains of our ancestors underwent rapid enlargement, we would recognize topographic features: cliffs, waterfalls, rivers, and lakes. The landscape we would need to traverse to seek food and find shelter would be familiar. We would recognize most species of plants and animals. The sounds of nature and their sources would be familiar. Scientists now think that many universal human traits, such as classification of basic colors, probably evolved in response to those stable parts of the environment.[11]

Other environmental features, however, have appeared so recently that we lack evolved responses to them. Some of them, such as domesticated plants and animals, exotic species, dams, and buildings are relatively minor modifications of long-standing environmental features. Others, such as highways, toxic chemicals, and machinery, are novel. They have been present for such a short time that we have not evolved adaptive responses to them. As we will discuss, our innate knowledge does not help us deal with them.

The same unconscious assessment happens with the people we encounter. Our social brains are wired to calculate the positive and negative affordances of people we meet. Later we will discuss why we are especially good at sizing up strangers and reading body and face language (figure 3.3).

Figure 3.3. By observing infants' brain activity in response to their mother's face versus that of a stranger, scientists have learned that, by at least the age of six months, infants are able to distinguish between familiar and unfamiliar faces.

Whatever their nature—the environment, the plants and animals it in, or the people with whom we share the space—conditions persist for only a particular length of time. Just how long influences when and how we need to respond to them.

Timing Matters

Most objects moving across the ancient landscape were potential prey, predators, or other people. Tracking them was important and decisions often had to be made quickly. In addition to responding quickly to events, our ancestors had to pay attention to environmental cues that change slowly over time. Seasonal changes were among the most important ones. The types of resources and where they could be found changed as the year progressed. Our ancestors had to gather and act on this information, for instance, deciding to follow migrating herds, or journeying to a higher elevation to harvest a particular wild food.

Some features of the environment or events changed minute to minute, or hour to hour. Others changed annually; still others on a scale too slow for us to register. Changes that occurred on the scale of minutes or hours included shifts in the weather, the sudden appearance of wild animals or potential enemies, and the approach of dusk. Responses to these near-term events included seeking shelter, taking defensive action, and finding a secure sleeping place before nightfall. Seasonal changes included shortened days; the appearance of buds, leaves, and flowers; rainfall. In response to these changes, our ancestors shifted their hunting sites. Changes on the scale of decades included alteration to the habitat as meadows were replaced by woodland or a river changed course. In response to these longer-term changes, small bands may have set out in search of new territory to settle. Features that changed on the scale of centuries would have registered as constants. An ability to anticipate the future and plan for it is vital for dealing with changing conditions.

Filtering Environmental Information

Much of the environmental information our brains receive is of little or no consequence. We can't pay attention to all of it, and we shouldn't. So how do our brains sort through the flood of data and winnow out only the information upon which survival depends?

Our nervous system does so in several ways. Sense organisms have evolved to respond only to parts of the incoming information. For example, some animals (birds, insects) can see ultraviolet light; we cannot. Also sense organs are especially sensitive to sudden changes. Constant environmental conditions are generally of little immediate significance. Sudden changes, on the other hand, are usually worth paying attention to (think of a rabbit's sudden swiveling of its ears in response to a sound). We respond weakly to background smells, but react strongly to sudden odors, like smoke from the toaster. In a more dramatic example of filtering, the ears of a bat are completely turned off while the bat is echolocating; the ear is then ready to respond to soft echoes when they arrive a fraction of a second later. If the ear were operating during calling, the call's loud clicks would still be reverberating in it; the bat could not detect the soft echoes that tell it where prey or obstacles are. The bat's survival depends on this ability to filter out an extraneous signal—its own echolocation call.

Estonian biologist Jakob von Uexküll was the first to recognize the importance of information filtering.[12] He thought that the subjective worlds of animals were made up of a small number of things important to it. He called these subjective worlds *umwelt* (German for "environment"). The *umwelt* of a tick, von Uexküll argued, consists of only three things: The odor of butyric acid, produced by the sweat glands of all mammals; the temperature of thirty-seven degrees Celsius (the temperature of mammalian blood); and the hair of mammals. By responding to those few cues, a tick can find and attach to a mammal, the primary task it needs to accomplish.

Our *umwelt* is more complex than a tick's, but we too often rely on simple cues. Those cues are largely visual. Primates, including humans, rely on vision to locate things, to identify objects from fragmentary information, and to detect when things move. As befits a diurnal primate we have forward-facing eyes and color vision, features that help us comprehend environments. To tell where objects are in space we need to perceive depth. Most of us use differences in the images falling on our two retinas to do so, but individuals with only one good eye still perceive depth. They manage this by using other cues: the relative sizes of objects, which objects hide others, shadows, elevation, texture gradients, color, and linear perspective.

Our brains also extract patterns from complex backgrounds and recognize them as objects. We construct complete figures from fragmentary cues (a snake in the grass) by having strong line and edge detectors combined with a neural program that interprets even poorly defined lines and incomplete objects as parts of real physical forms. Our color vision also helps us find objects. Studies of infants show how innate information helps our children cope with the complex world into which they are born. The visual systems of children mature slowly, but even as infants our brains actively filter visual information. Later in the book, we will look in detail at how the relevance of environmental information changes with age and why infants' innate knowledge is adaptive.[13]

As we have just discussed, animal minds possess innate knowledge about the environment because individuals that possess internal models of how the world works make better decisions than individuals that lack them. An evolutionary view explains the fit between innate human knowledge and our environment. It also explains why we have a central nervous system that stores past knowledge that served our ancestors.

So our ghosts of the savanna are evolved responses to an environmental problem or opportunity that persisted long enough for significant genetic changes to occur in our species. I call them ghosts because many of the processes we use to assess the affordance of an environment are unconscious; we generally remain unaware that our nervous system is processing and evaluating the information we use to make quick decisions. Environmental information flows over and around us; we ignore much of it. But the relevant information is allowed through, allowing our unconscious assessment of our surroundings.

With the background provided by these three chapters, we are now ready explore in detail how we interact with the environment, to explain why our ancestors' emotional responses helped them respond appropriately to environmental challenges, and to determine why the consequences of their responses reside as ghosts of environments past in our minds today.

4

Reading the Landscape

Ghosts of the African savannas inhabited our ancestor's minds as they dispersed from Africa to Europe and Asia. Archaeologists have learned this because several kinds of information tell them what kinds of environments our ancestors traversed and the resources they used. For example, traces of plants and animals in their caves and campsites reveal what they ate. The remains of more than 250 species of birds have been found in campsites scattered over much of Europe. Those birds exist and use the same habitats today as they did thousands of years ago, so archaeologists use them to infer the landscapes around the caves. Those birds, and, hence, our ancestors, lived in habitats, like the East African ones, that were a mixture of savanna, wetlands, and rocky outcrops.[1]

Our ancestors' artworks also tell us what they paid attention to. An engraved stone tablet was found in a cave in Abauntz in the Navarra region of northern Spain (figure 4.1). It was carved about 13,600 years ago and contains the earliest known representation of a landscape. The stone measures less than five inches by seven inches and is less than an inch thick, but it depicts mountains, meandering rivers, ponds, and good foraging and hunting places (figure 4.2). Access routes to different parts of the landscape are carved on it. A mountain with herds of ibex on its slopes is clearly one that can be seen from the cave today. The landscape is strikingly similar to the East African home of our ancestors—a mosaic of open grassland, widely spaced trees, scattered copses, and denser woods near rivers and lakes. Grasslands with scattered trees, hills, and rock outcroppings are an ideal habitat in which to hunt. The inhabitants of Abauntz Cave could have used the wide vista to plan long-distance moves; the trees and prominences were places from which they could

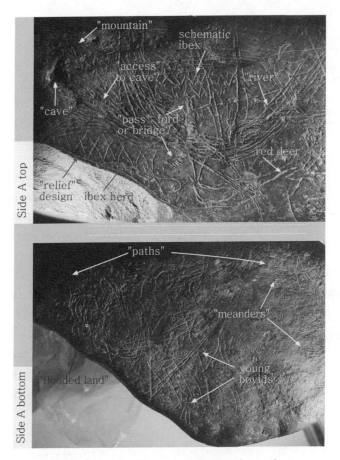

Figure 4.1. A primitive map engraved fourteen thousand years ago on a rock by Magdalenian hunter-gatherers in what is now Spain. Pilar Utrilla of University of Zaragoza discovered the rock in the cave of Abauntz Lamizulo in 1994.

track moving animals and detect approaching human groups. Nearby lakes and rivers were permanent sources of water.

In combination, these results also suggest that our ancestors probably dispersed along coastlines rather than via the large treeless areas of interior Europe. Coastlines offered the mixture of savanna, woodland, outcrops, and water that our ancestors needed. They were not yet able to kill large animals at a distance. That skill would have been essential on the open plains that lacked concealment.

Ghosts of the African savanna accompany travelers today. For example, in September 1849, Captain R. B. Marcy was leading an expedition to explore little known areas of the southern plains for the American government. He recorded the following emotions as they approached the headwaters of the Clear Fork of the Brazos River "over as beautiful a country for eight miles as I ever beheld. It was a perfectly level grassy glade, and covered with a growth of large mesquite trees at uniform distances, standing with great regularity, and presenting more the

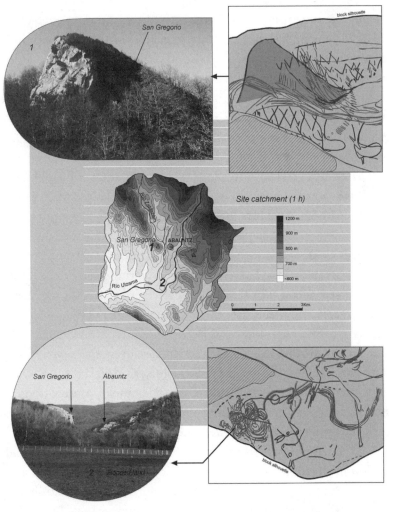

Figure 4.2. A representation of a map on the rock from the Late Magdalenian level from Abauntz Cave. *Top,* detail of the actual mountain and its representation on the rock, with the river and the ford at its feet. *Bottom,* landscape features identified by numbers.

appearance of an immense peach orchard than a wilderness." Captain Marcy intuitively inferred that the savanna landscape was fertile, well watered, and would be a good place to settle. His responses, like those of the rest of us, reflect our long time on African savannas.[2]

We seem to have retained these emotional responses to landscapes for thousands of years. They are ghosts of past environments that reside in our minds. To gain insight into how such behaviors can become part of a species' DNA, let's imagine an animal first entering an unfamiliar environment. Of the data flooding its senses, which bits of information should it pay attention to? Which ones would tell it the most about what this environment has to offer? Current conditions matter, but if the individual is to claim the area and raise offspring there, future conditions may matter even more. For this reason, many animals attend to those features of the landscape that predict future conditions. These features may not have anything to offer the individual over the short term. For example, birds may settle not where there is an immediate abundance of food, but where there is promise of future food. Birds use features such as the density of stands of trees and the configuration of their branches and leaves when choosing a nesting site.[3] Birds evolved to do this because vegetation structure is a better predictor of food supplies weeks later, during the crucial period when hungry nestlings will need to be fed. Birds assess not only what the habitat can do for them now. They also assess what the habitat has to offer their offspring weeks in the future.

So we see that an animal selecting a place to live makes a sequence of decisions. First, it selects the general area, one with a variety of habitats. Different habitats are suited to different activities. In one, an individual may display to other males and court females. In another, it may forage. It may select a nest site and defend it against rivals in a third. Finally, it must decide how to respond to individual features of the environment as it encounters them. Depending on the species, these can be potential predators, prey, mates, rivals, nest sites, and so forth.

Individuals typically make decisions about individual features of the environment (eat those berries) more frequently than ones that affect behavior for a longer time (build a nest here), but the latter strongly affect the others. For example, a bad decision about a single food item—rejecting it when you should have eaten it—will affect survival much less than a bad decision about where to place a nest, which will influence hundreds of foraging decisions. However, some decisions about individual features, such as a slow response to a hungry predator, may have lethal consequences. In general, the more time and energy an individual invests in a decision, the more important that decision should be to its survival and reproductive success. This holds true for humans as well. We invest more time choosing a house than deciding on a restaurant for dinner; courtship often extends for many years.

An individual's status—age, sex, and physical condition—also influence where it goes, how it gets there, and what it does when it arrives. Time of day and weather also influence its decisions. If the decisions help the animal survive and reproduce, that is, if a safe nest site in an area rich in food is chosen, the psychological mechanisms that influence such decisions evolve to become encoded in the animal's genes.[4]

Stages of Habitat Selection

My colleagues and I have found it useful to think of habitat selection as a three-stage process.[5] We'll call these stages encounter, explore, and exit or establish. In the encounter stage, an animal comes upon an unfamiliar landscape and decides either to explore it or to move on. As we will see, people make this decision unconsciously and almost instantly. Decisions at this stage do not require conscious inference, but the individual may unconsciously consider information he or she has stored about places he or she has already been.

These rapid, unconscious responses are evoked by parts of the environment that signal its suitability. Typically, these features are general ones; psychologist Robert Zajonc called them "preferenda."[6] They include the distribution of objects in space, depth clues, water, and trees.[7] We use clues about depth to assess distance, perhaps to estimate the time required to cross open spaces to reach prey or a place of safety. Trees and flowing water signal the presence of food, possible shelter, and the ability to move unimpeded by following the course of a river.

If the individual decides to remain, it begins to explore the area and gathers more information about it. This explore stage involves a great deal of cognition; exploration may extend over many days. Complex, patterned features of the environment, such as views of the horizon and ways of moving through it, prompt exploration. Only by exploring a complex environment can an individual determine whether resources are abundant or scarce, now and in the future, and whether it offers safety or peril.[8] Exploration is risky, exposing the individual to as yet unknown danger, so risk assessment is a prominent component at this stage.

British geographer Jay Appleton developed a theory to explain how we evaluate unfamiliar landscapes by considering features—prospect, refuge, and hazard—that influence our ability to safely explore and gather information about a new environment.[9] *Prospect* offers a good view or vista (figure 4.3). *Refuge* provides cover from which to observe without being seen. *Hazard* refers to the risks a person faces while exploring an environment. These features could be primary—available immediately from the viewer's current position—or secondary, accessed only after entering the environment. Elements of prospect-refuge theory are relevant to both our unconscious initial responses to unfamiliar landscapes and the conscious assessments we make while exploring an area.

Appleton conceived this theory, which has stimulated much research

Figure 4.3. From a perch in a tree high on a ridge, a Hadza man named Mahiya peers across the rough terrain, looking for game.

on landscapes, while reading a paragraph in Konrad Lorenz's classic work of popular science, *King Solomon's Ring:*

> Easter is already in the air, and we are taking a walk in the forest whose wooded slopes of tall beeches can be equaled in beauty by few and surpassed by none. We approach a forest glade. . . . Before we break through the last bushes and out of cover on to the free expanse of the meadow, we do what all wild animals and all good naturalists, wild boars, leopards, hunters, and zoologists would do under similar circumstances: we reconnoiter, seeking before we leave our cover, to gain from it the advantage which it can offer alike to hunter and hunted—namely to see without being seen.[10]

Appleton's ideas have influenced many workers across diverse disciplines—my own included. We will revisit prospect-refuge theory over the next three chapters of this book.

During the third and final stage, the individual moves on (exit) or decides to remain, for a short time, a season, or a lifetime (establish). If it elects to stay, an individual may try to improve the environment (excavate a burrow, build a nest, clear brush, dig a well). The stages of habitat selection often overlap; so will our discussions.

Our responses to environmental cues vary with our needs. Food, water, and shelter matter more when we are hungry, thirsty, or cold. A family enjoying a picnic is likely to dislike an approaching storm; a farmer whose crops are wilting in a drought will welcome it. But they will all seek a dry place before rain falls. Despite these differences, common patterns emerge because humans respond positively to evidence of food, water, shelter, and refuges, and negatively to potential hazards, such as storms, fire, predators, and obstacles.

The Legible Landscape

In the developed world, we spend most of our lives in familiar environments—our homes, workplaces, areas where we shop. We function well in those environments and generally feel comfortable there. Yet even the most settled among us regularly find ourselves in unfamiliar places. Our ancestors made many first-stage decisions during their lives. We can infer how they responded by studying how we comprehend unfamiliar environments today.

Much of what we know about our unconscious responses to environments comes from experiments with images. A rich experimental literature shows that people quickly and unconsciously draw inferences when shown photographs of unfamiliar landscapes.[11] We base these inferences on how much information we need to process to draw inferences about the scene before us. We draw some of these inferences quickly and unconsciously (encounter stage), but we may then employ substantial conscious processing (explore stage).

Psychologists use terms such as "coherence," "complexity," "legibility," and "mystery" when talking about how we unconsciously and consciously process a landscape when we first encounter it.

Coherence refers to the ease with which we can grasp a scene's organization. *Legibility* is an assessment of how well one could find one's way in it without getting lost. American urban planner Kevin Lynch first introduced the concept of legibility in his 1960 book *The Image of the City*.[12] He described the apparent clarity or legibility of a cityscape in terms of the ease with which we can recognize its components and organize them into a coherent pattern. A legible space is one that is easy to understand and remember. It has a structure that enables us to navigate within it and find our way back to the starting point.

To assess *complexity* we pay attention to objects that may tell us about the scene's resource richness. *Mystery* refers to a promise of more information if we enter the scene. Mystery arouses our curiosity and evokes hypotheses about where to find resources and where risks and dangers are most likely to confront us. But to gain that knowledge, we must expose ourselves to danger. As Konrad Lorenz suggested, it is advantageous to see without being seen.

As we will see, our responses to paintings and photographs of landscapes are a rich source of information about how we evaluate unfamiliar landscapes. Experiments conducted by many investigators reveal regular and meaningful patterns to our rapid evaluations, but that subjects are generally unable to explain their choices. Even though we like to think that we understand why we respond the way we do, experimenters have found that when they ask us why we like or dislike scenes, we are often embarrassed that we cannot explain why;[13] we invent explanations. Psychologist Jonathan Haidt wrote in his 2006 book,

When you see a painting, you usually know instantly and automatically whether you like it. If someone asks you to explain your judgment, you confabulate. You don't really know why you think something is beautiful, but your interpreter module is skilled at making up reasons. . . . You search for a plausible reason for liking the painting, and you latch on to the first reason that makes sense (maybe something vague about color, or light, or the reflection of the painter in the clown's shiny nose).[14]

Fortunately, as we will explore in later chapters, researchers have devised clever experiments to help reveal our hidden motivations.

The features that influence our initial, noncognitive responses to unfamiliar environments in the encounter phase are also important during the explore phase. Ghosts of environments past also influence how we assess an environment as we explore it, respond to its risks and resources, and alter it to improve our lives.

These assessments are the focus of the next several chapters.

5

The Snake in the Grass (...and Other Hazards)

While I scaled the trunk of an acacia tree, my companion stood nearby, scanning the horizon through binoculars, keeping a lookout for predators. I was climbing the tree to measure the height at which the trunk split into main branches and to collect leaves to study later in camp. It was October 12, 1978. My wife, Betty, and I were camped at the base of the Nguruman Escarpment in southern Kenya where we were gathering data to test a prediction.

Earlier that year I had proposed the savanna hypothesis, and we had come to Africa to put the hypothesis to a test. When I advanced the savanna hypothesis, anthropologists already knew that our ancestors lived in East African savannas during the time their brains tripled in size. I predicted that the shapes of trees that dominated resource-rich savannas in which they would have survived and reproduced best should be especially attractive to us. So Betty and I measured trees growing on resource-rich and resource-poor Kenyan savannas to find out which shapes dominated the best sites. But we had more on our minds than the shapes of the trees—the shapes *among* the trees and in the tall grass concerned us even more. At night in our tent, we could hear lions roaring; by day, as we walked across the savanna, we wondered how close to our path a snake might be resting.

We were worried about lions, but Betty and I were especially concerned about snakes. What if we failed to spot an adder or a large python until we got within striking distance of it? Startling at the sight of a snake or even a shed snakeskin is a nearly universal human response. Snakes evoke disgust in many people who mistakenly view them as "slimy" because they glisten.[1] Snakes evoke strong negative emotions in people who have never seen or been bitten by one. New Englanders, few of whom said that they had previous encounters with snakes, reported that snakes were the most prevalent objects of their intense fears.[2] But is our fear of snakes innate, a legacy of our primate ancestors? Or is it learned?

The neural systems that prompt us to flee from or avoid danger were probably among the first behavioral systems to evolve in animals.[3] To stay out of trouble, an animal must be able to recognize patterns, know which ones signal danger, and act appropriately. The objects and situations that evoke fear in humans today—snakes, large mammals, hostile strangers—were genuinely dangerous to our ancestors. A fear-and-learning neural circuit shared by many mammals, including ourselves, is located in the amygdala, an almond-shaped region of nerve cells in the temporal lobes of the vertebrate brain.[4] This "fear circuit" yields rapid responses that can be triggered by very few clues.[5] It is especially sensitive to stimuli, something half-seen, undulating through the grass, that are related to recurrent survival threats.

Venomous snakes and large constrictors like boas and pythons have caused human deaths for millennia.[6] They preyed on our primate ancestors, on the earliest hominids, and for hundreds of thousands of years, they have preyed on us. Snakes have a highly distinctive shape and way of moving. Most snakes are ambush predators; they are dangerous only if we approach them closely without detecting them. Most other large terrestrial predators also need to approach their prey closely before they launch an attack. They take advantage of cover and have color patterns that camouflage them against the backgrounds in which they hunt.

Our ancestors would have benefited from an ability to detect hidden, stationary predators from glimpses of only small parts of them. Special systems for distinguishing the patterns that provide camouflage—stripes, spots, complex patterns with crossing lines and many edges (tessellated patterns)—and for detecting their eyes would help. Research suggests that we do have these special detection systems.

Tessellated patterns are rare in nature but common among snakes.[7] Yet, cells of the mammalian visual system are highly stimulated by such patterns. The system readily detects scale patterns in peripheral vision, where we are most likely to spot a snake. A checkerboard arrangement of textures presented close to where a person is looking generates more activity in our visual centers than a uniform texture.[8] A diamond pattern produces larger neural activity in the visual centers of the human brain than a random-dot pattern, a circle, or a triangle on a randomly textured background.[9] The neural system responds selectively to angles and edges, features that enhance our ability to detect snakes against a background during daylight. Our neural fear module increases our ability to detect motionless snakes, even when we are unaware that we have seen them.[10] Lynne Isbell makes a compelling argument that enhanced ability to detect motionless snakes was the factor that favored the evolution of the unusual features of the primate visual system.[11]

Yet, reptilian scales also evoke positive emotional responses. During the last decade of the twentieth century, snake prints were fashionable in women's apparel and accessories. They were worn to gain attention. Skins of crocodiles and snakes were often used in consumer products, although today, in response to conservation concerns, printed fabrics that resemble real skins replace them. In south Asia, Hindus, hoping to enhance their fertility, worship cobras. They associate cobras with the sacred symbol of Lord Shiva. During a serpent ceremony, dense crowds croon and pray to each cobra displayed before the deity (figure 5.1).[12]

Figure 5.1. People offer prayers to a snake during the annual Hindu Nag Panchami festival in Jammu, India.

Figure 5.2. *Left*, scales of a gaboon viper (*Bitis gabonica*); *right*, detail of the façade of Shir Dar funerary madrasa, Samarqand, 1636.

Our desire to approach and investigate snakes probably accounts for the frequent appearance of scale patterns on engravings and their display in paintings and mosaics. Deliberately ground, scraped, and gouged shale on which ocher pigment was engraved with a crosshatch pattern similar to that of reptiles are among the oldest artifacts of Middle Stone Age sites. The dominant diamond-patterned crosshatching, zigzags, and step forms of the decorative motifs in Mayan art are thought to imitate rattlesnake scale patterns.[13] Some of the mosaic designs in Islamic art also are similar to snake-scale patterns (figure 5.2).

During the early evolutionary history of primates, venomous and constricting snakes were their most important vertebrate predators.[14]

Dangerous snakes continue to be a source of human disability and death. More than 150,000 people die from snakebites each year, most of them in the tropics.[15] Other than lemurs, which evolved on Madagascar, an island lacking venomous snakes, all primates have an innately based fear of snakes that is easily triggered. Laboratory-reared monkeys that have never seen a snake become fearful merely by observing other fearful monkeys.[16]

We also have a built-in bias toward detection of and responses to snakes. Nine- to ten-month-old human infants do not react differently to snakes than to other animals, but by the age of one year, before they begin to walk, infants quickly associate snakes with fear. By the age of three, children detect snakes more rapidly than nonthreatening stimuli (flowers, frogs, caterpillars). In contrast, neither children nor adults detect a single frog among flowers better than a single flower among frogs.[17] In other words, we learn more quickly to spot and respond to snakes than to most other things because neural circuits that respond especially strongly to them form during the first year, a time of rapid maturation of the infant brain.

A by-product of our predisposition to fear snakes is that we mistakenly infer correlations between snakes and aversive stimuli. In an experiment, Andrew Tomarken and colleagues showed study subjects slides of snakes, flowers, or mushrooms. Some slides were paired with a slight electric shock; others were not. The study subjects were more likely to mistakenly judge that slides of snakes had been paired with shock than were slides of flowers or mushrooms.[18] They did not experience an illusory shock when shown dangerous man-made objects such as damaged electrical equipment or guns. The image of a snake is sufficient to make us imagine we feel pain.

Given that a snake poses relatively little danger once we are aware of it, a certain level of attraction to snakelike forms would remind us of its presence and induce us to monitor where it is. Evidence from fieldwork with primates suggests how this might have evolved. African and Asian monkeys and apes call to draw attention to a snake. The alerted group members may follow the snake until it leaves the area to hunt in a place where it has not been seen.

The mixture of fear and attraction we have to snakes probably explains why snakes have become symbols of power and sex, totems, protagonists of myths, and gods. Snakes are mystically transfigured in most cultures. The Hopi recognize the water serpent Palulukon, a benevolent but frightening godlike being. The Kwakiutl, who live in coastal British Columbia and Alaska, a region that lacks venomous snakes, fear *sisiutl*, a three-headed serpent with both human and reptilian faces. When it appears in dreams, it presages insanity or death. The Sharanahua of Peru summon reptile spirits by taking hallucinogenic drugs, and they stroke their faces with snake tongues.

Awe of snakes has generated much art, especially religious. Serpentine forms adorn European Paleolithic stone carvings. Snakes were carved into mammoth teeth in Siberia, well outside the range of any snake. Stylized snakes often serve as talismans of the gods and spirits that bestow fertility: Ashtoreth of the Canaanites, the demons Fu-Hsi and Nu-kua of the Han Chinese, and the powerful goddesses Mudammã and Manasã of Hindu India. Ancient Egyptians venerated at least thirteen snake deities that ministered to health, fecundity, and vegetation. Amulets in gold inscribed with the sign of a cobra god were placed in the wrappings of Tutankhamen's mummy. The Aztec rain god Tlaloc consisted in part of two coiled rattlesnakes whose heads met to form his upper lip. Coatl, the serpent, is part of the names of many Aztec divinities. Coatlicue was a threatening chimera of snake and human parts; Cihuacoatl was goddess of childbirth and mother of the human race. Fire was rekindled every fifty-two years over the body of Xiuhcoatl, the fire serpent, to mark a major division in the Aztec religious calendar. Quetzalcoatl, a plumed serpent with a human head, was god of morning and evening, of death and resurrection (figure 5.3). Aztecs believed that he invented the calendar and writing.

Snakes evoke strong fearful responses in most people, but we are also afraid of many other plants and animals. Injury and death caused by other organisms have been important throughout human history, but responding to other organisms is complex because we get many positive things from them as well. Let's look more closely at the threats many other animals and plants pose to us to understand why we might fear them.

Dangers Large and Small

Plants and animals provide us with food, fiber, fuel, transport, and protection, but some of them may be dangerous. Plants remain rooted to

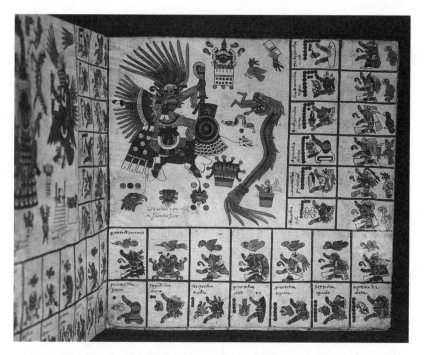

Figure 5.3. The gods Tezcatlipoca and Quetzalcoatl devouring a man. "Tonalamatl" (Codex Borbonicus, religious calendar), Aztec manuscript on paper, 39 × 40 cm, early sixteenth century, Bibliotheque de l'Assemblee Nationale, Paris.

one spot and, unless we ingest them, pose little threat. The threats posed by other animals are many. They are mobile, often stealthy, and use camouflage to avoid being seen. Animals can be predators or parasites, and even when we ourselves are the predator, contaminated meat can sicken us. The threat posed by an animal may depend on its current state. We need to worry about a hungry lion but not one that has just fed on a zebra.

Dangerous organisms can range in size from elephants and large constricting snakes to scorpions and spiders, to the unseen, invisible to the unaided eye, disease-causing bacteria, fungi, and viruses. Animals that otherwise do not harm us—flies, fleas, ticks, and mosquitoes—can transmit disease. We did not know of the existence of microorganisms until the seventeenth century, and we did not know that they cause diseases until only a century ago, so we have not evolved to fear them. We do, however, experience revulsion when we see things that are associated with pathogens—festering wounds, rotting meat, dead bodies, and feces.

Thorns, Horns, and Things with Sharp Teeth

Predators large enough to pose a danger to a full-grown human have large, pointed teeth and claws. Many large grazing mammals have sharp horns, antlers, or hooves. They use their horns in sexual combat; they would have used them against attacking human hunters. Many species have facial patterns that exaggerate the size of their weapons. Some plants have sharp, penetrating thorns and spines. We should be motivated to avoid contact with them but should know and remember where they are. As with other potentially dangerous things, we may be both fearful of and attracted to them. We have a deep fear of long, sharp teeth. Males of many primates have prominent canines they use to threaten and fight enemies and competitors. The Dracula legend continues to inspire fear even though males of our ancestors long ago lost their protruding canines.

Responses to Pointed Forms

We have strong responses, both positive and negative, to pointed forms. African savannas are full of plants with sharp thorns. Most species of *Commiphora* and *Acacia*, which dominate vast areas of African savannas, have spines that can puncture our skin, feet, and eyes. Paying attention to those plants, especially during times of rapid movement, would have reduced the chance of injury.

Richard Coss videotaped walkers and joggers as they passed two plants positioned on each side of a path in the arboretum of the University of California, Davis. The "provocative" plant, a Spanish dagger (*Yucca gloriosa*), has long, dagger-shaped leaves that radiate from a central stem. The "safe" plant was a similar-sized crape myrtle (*Lagerstroemia indica*) with rounded leaves. Coss switched locations of the plants halfway through the experiment. We might have expected them to avoid the dagger, but both walkers and joggers favored the side of the path where the dagger was located; they passed close to it.[19] Coss explained this surprising result by suggesting that paying attention to the dagger might help joggers remember its location and be less likely to run into it in the future.

Early humans learned to fracture rocks to create flakes and cores they used for cutting, scraping, and tips of weapons. These pointed objects were valuable, so our ancestors should have evolved to find them attractive. Coss tested this prediction by recording people's responses to black silhouettes of pointed and rounded shapes. People judged pointed shapes to be both more dangerous and more attractive than blunt shapes.

Because pointed shapes are aesthetically attractive but also evoke caution and fear, we expect sharp forms to be prominent in art and design. They are! Pointed forms are common in sculptures, graphical designs, and shrines. They are used to increase the aggressive appearances of characters in plays and dances. The gaping mouths and prominent teeth of figures that adorn temples and medieval cathedrals were thought to have warded off evil.

Another reason for our strong responses to pointed forms could be that the angles in the eyebrows, cheeks, chin, and jaw in an angry human face resemble a downward pointing V-shape. The curves in the cheeks, eyes, and mouth of a happy face are round.[20] A simple downward pointing V triggers greater activation of several brain regions than does an identical V-shape pointing upward. We detect downward pointing V-shapes embedded in a field of other shapes more readily than we detect the same shapes pointing upward.[21] Only a person adopting an evolutionary approach would have thought to test this prediction or have imagined the surprising result.

The Leopard's Spots

Large spotted cats have preyed on primates for millennia. Even today leopards cause many human deaths in Africa and the Indian subcontinent.[22] Leopard rosettes fascinate infants only seven months old. In a series of experiments in day-care facilities, infants and toddlers were allowed to handle four lightweight plastic jars. Each jar had yellowish-orange paper with full-sized patterns of leopard rosettes, rock python patches, or scales pressed against its interior side and bottom surfaces. Children poked jars with python and leopard patterns significantly more often than they poked plain and plaid jars. Richard Coss interpreted poking as an investigative action because children immediately stopped touching jars if they rolled over.[23]

Benefits from paying attention to large cats may account for our fascination with and attraction to their fur. The first recorded use of spotted leopard pelts as clothing comes from early Neolithic hunters and individuals of high stature in the village of Çatal Hüyük.[24] Some societies still associate wearing spotted pelts with high social status.[25] The value of spotted-cat pelts today is so great—as much as $60,000 for a single snow leopard pelt—that illegal poaching threatens some species.

Eyes—Has It Seen You?

Most prey animals have eyes on the sides of their heads that enable them to detect predators sneaking up from behind. The eyes of most predators, on the other hand, are on the front of their heads. Forward-facing eyes enable them to more easily find and track their prey.

In study after study, in species after species, scientists have found that many animals pay special attention to eyes and eyelike forms. Macaques have a special neural system that facilitates recognition of two facing eyes.[26] Wild bonnet macaques recognize leopards better if they can see their facing eyes.[27] Newborn human infants respond to a pair of large facing eyes by turning their heads. Even infants lacking head control react by withdrawing; preterm infants as young as thirty-six weeks old react to eye contact with strangers by stiffening their bodies and averting their eyes.[28]

We most easily detect a vertebrate, particularly a mammal, and assess its intentions by looking at its face.[29] We also respond to eyelike symbols, such as bull's-eyes. Automobile taillights designed as two concentric disks elicit much stronger physiological arousal (skin conductance, pupil dilation) than other taillight patterns.[30]

Friend or Foe? Assessing the Intentions of Other Humans

Throughout human history, we have distrusted people outside our clan, our tribe. Even though research shows that most violence today comes at the hands of people we know, and even though we are all descended from a common ancestor who lived some two hundred thousand years ago, we still fear people we perceive as "other." We believe strangers will do us harm.[31] Why? In deep time, hominids outside the family group were likely to be a raiding party.

Fear of strangers appears among children as soon as they can move on their own.[32] To assess the intentions of other people, we look at all body parts, but our faces are especially revealing. The human face has unusually complex musculature and neural innervations. Facial muscles differ from most other muscles by moving skin rather than bones. About twenty muscles produce psychologically meaningful facial expressions.[33] Darwin clearly recognized that the design of the human face suggested that it had evolved to communicate social signals.[34] Facial expressions of most primates are similar to ours.[35]

As we have already seen, signals can evolve only if the behavior they elicit benefits both signalers and recipients. Facial expressions meet these requirements. We all gain from an ability to detect subtle facial clues. We may also gain by giving false information about our future behavior. Angry males pose greater danger than angry females. We do not know how often males attacked infants during our evolutionary past, but males of many mammals attack infants.[36] Most attacks happen when strange males evict the dominant resident males of a social group. By killing infants they induce lactating females to come back into estrous quickly. Fear of strangers doubtless has deep roots in human evolution.[37]

Therefore, men should react more strongly to angry men than to angry women. Experiments confirm this expectation. Men show stronger physiological (skin conductance) responses to angry men than to angry women.[38] Infants also fear men more than women.[39] These results are not caused by height differences because tall women do not elicit as much fear as men do. Nor can they be explained by differences in facial hair; infants whose fathers had beards were just as afraid of strange men with beards as infants with beardless fathers.[40]

Fear of strangers develops around seven months, peaks at about one year, and continues until two years of age. Fear of strangers does not develop until several months after infants discriminate between familiar and unfamiliar people.[41] Until they start crawling infants are unlikely to contact strangers without being with their mothers. This developmental "gap"[42] makes sense from an ecological and evolutionary perspective.

A Dangerous World

Many physical hazards are permanent features of landscapes. To avoid injury or worse, we need to remember the location of cliffs, waterfalls, rapids, and other dangerous features. An evolutionary perspective suggests that people should be motivated to learn and remember their locations. Our ancestors doubtless kept such knowledge in a mental map and used it to plan future hunting and foraging expeditions or war parties. We should be motivated to approach closely enough to see how dangerous they are, whether they offer any reward despite the danger, and how best we should respond to them. This may explain our fascination with waterfalls and cliffs and why we enjoy approaching them, but not too closely. Most national parks in the United States were established around spectacular but potentially dangerous geological features (figure 5.4).

It seems our brains may even be wired to exaggerate the danger, just in case we don't get the message.[43] We respond strongly and rapidly to some hazards because we have an inflated perception of the danger.[44] Consider the danger posed by uneven ground. We may fall when we move over an

Figure 5.4. Denali National Park, Alaska, is dominated
by North America's highest mountain.

uneven area with vertical changes less than a meter. A simple trip can result in an injury that makes it harder to walk; falls of only a meter or two can produce serious injuries. These risks would have been especially important when our ancestors were pursuing prey or escaping from enemies, or even when they had to cover a lot of ground while foraging to meet their energy needs. Even if a fall were not fatal, it could have serious consequences, leaving the injured vulnerable to predators or simply unable to find enough food. As we would expect, our perceptual system makes vertical surface irregularities seem larger than they are.[45]

We should also pay attention to signals that something dramatic is about to happen: a loud noise, a strong vibration, or a flash of light. Loud noises accompany or immediately precede many natural disasters and other hazardous events. We pay attention to the sound of thunder and falling trees or rock slides. Fear of loud noises and bright lights (associated with thunder and lightning) appears during infancy.[46] By the time they are six years old, children fear more events—earthquakes, fire, thunder, lightning, and deep water.[47] Such events should command our immediate attention because the time to initiate a timely response may be very brief. However, cognition is usually necessary to determine the cause of a noise, its location, and to decide on an appropriate response to it.

Water Hazards

Of all the world's physical features, water is the most changeable. Tsunamis and floods can strike without notice. People can quickly drown in ocean waves and riptides. Tides can cut off access to land and strand us. Even permanent features, such as waterfalls, rapids, and deep lakes can be hazardous.

We have many reasons to fear water, but we need to drink daily, and lakes and rivers are rich sources of food, so we should also be attracted to water. As we will see in chapter 6, we are! We descend on rivers, lakes, and oceans for our vacations. Landscape architects go to great lengths to enhance water in public parks and gardens. We pay more for homes that front on or have views of water.

The result of our ambivalent responses to water is that we are attracted to water but we keep a safe distance from dangerous features.

We get close enough to waterfalls to view them without incurring a significant risk.

Light and Shadows

Night was a dangerous time for our ancestors and other diurnal primates. Fear of darkness must have evolved soon after nocturnal primates became active during the day. Infant baboons are born with fears of falling and the dark. Our visual system, evolved to see color during the day, functions poorly under low light. Many dangerous predators (hyenas, large cats, wild dogs, and snakes) hunt primarily at night. Our ancestors, who daily covered long distances hunting and gathering, may often have found themselves without shelter at dusk. The setting sun would have been a strong signal to return to a safe place before nightfall. Evening shadows also provide a brief period during which it is easier to perceive depth than at other times during the day.

Therefore, cues that signal the approaching darkness—setting sun and lengthening shadows—should be highly motivating. They are, but how we respond to a sunset depends on whether we are close to or in a safe resting place or in an exposed situation far from safety. Sunrise rarely requires urgent responses. Normally we view the rising sun, if we are awake, from a refuge where we spent the night. Light increases at sunrise, signaling many hours of good visibility.

Artists have apparently intuitively understood this, because they position people differently in sunsets and sunrises. In paintings designed to evoke pleasant feelings, people are closer to a refuge in paintings of sunsets than in paintings of sunrises. To elicit feelings of tranquility at sunset, landscape painters typically include an obvious and easily reached refuge. Pathways to the refuge are usually highlighted. Artists may even violate physical laws to highlight refuges: the sun's rays may be reflected from windows after the sun has set.

Fire: A Mysterious Hazard

Fire must have seemed mysterious to our ancestors. They probably knew that lightning caused fires (lightning-caused fires were common in the environments our ancestors inhabited), but they could not have understood why things burned or what happened to them as they burned.

Myths inevitably developed around something as mysterious as fire. For the ancient Greeks, for example, fire was stolen from the gods by the Titan Prometheus and delivered to Earth in a stalk of fennel.

Mastery of fire was a key event in human history, but to do so our ancestors had to overcome their fear of fire. They probably noticed that grazing mammals were attracted to recently burned areas to feast on the nutritious young blades of grass. They may have eaten and found tasty the scorched bodies of animals they found when exploring recently burned areas. Archaeologists have found evidence of our ancestors' wide range of uses of fire to influence where fresh young nutritious grass would be, to protect themselves from predators, to stay warm at night, to prepare skins, make pottery, and cook food. Hominids domesticated fire at least three hundred thousand years ago, perhaps as much as a million years ago. Fire became the most important tool our ancestors used to manipulate the environment. As people colonized tropical highlands and temperate lands, they carried fire with them.

Fears Change with Age

As we discussed in chapter 2, we are born with very immature brains. Brain growth that is completed during pregnancy in gorillas and chimpanzees continues for a year after birth in human infants. And genetically guided development of neural circuits continues for more than a decade. Methyl groups are added to DNA bases throughout our lives. They influence which genes are turned on and off as we mature.[48]

So genetically influenced behaviors appear at different ages. Genes govern our ability to speak a language, but we do not expect children to speak when they are born. Young women and men do not become sexually motivated and develop adult physical traits (breasts, buttocks, beards, low voices) until they are more than ten year old.

Although they are less obvious, children's responses to the environment also show evidence of adaptations that emerge over time as they become more mobile and encounter new spaces, organisms, objects, and situations that present both opportunities and dangers. As they age, children are increasingly likely to be farther from adults when they encounter hazardous situations. Children old enough to explore the world around them must increasingly rely on their own behavior; they can count less on protection and help from adults. For our hominid ancestors, cries or distress calls worked well only when a caretaker was within arm's reach, otherwise a cry might have attracted a predator. Natural selection should have resulted in neural programs that cause fearful responses to develop to specific things at ages when those things begin to pose a danger. We expect genetically influenced behaviors to first appear at the time during maturation that they would have been useful to our ancestors' children. The unfolding of our fear responses accords with this expectation.

Newborn infants cry or show other signs of distress from loud sounds, bright light, rapid or irregular movements, looming objects, and loss of equilibrium. Infants have good visual acuity by six to eight months. They point to many distant objects (cars, birds, planes). Between the ages of one and three years, children spend most waking time interacting with nearby objects,[49] a behavior that may reduce their tendency to wander.

Looming objects and sudden movements are likely to have accompanied the approach of dangerous animals.[50] Fear responses to these stimuli begin at about seven months. Fear of spiders begins around the age of three and a half years and continues throughout childhood.[51] Fear of "bugs" is greater among children six to eight years of age than among those aged nine to twelve.[52] Fear of small animals develops in children soon after they are capable of independent movement, the first time they are likely to encounter such animals when away from their mothers.[53] Fear of larger animals generally does not develop until children are more than four years old.[54] But fear of dogs has been observed in children as young as two.[55] After about ten years of age, social fears become more prominent; fear of large animals wanes.[56]

Being fearful of some objects and situations is rational. Indeed, as we have seen, most organisms that evoke fear were genuinely dangerous to our ancestors. But many people hold deep fears of situations, even imaginary ones that pose no danger. We call these apparently irrational fears "phobias." Psychologists recognize many different kinds of phobias: *nature phobias, animal phobias,* and *social phobias.*[57] An evolutionary perspective helps explain why we have these apparently irrational fears.

When Fear Becomes Phobia

We can make two kinds of mistakes when we sense danger. We can fail to respond when the danger is real, or we can assume that the danger is real when it is not. The costs of making these two kinds of errors differ dramatically. A false negative (failing to respond to a dangerous situation—a lion in the grass) is more costly than a false positive (overreacting to a situation that turns out to be harmless—wind whistling in the grass). The first error could be deadly; the second generally costs only wasting a bit of time and energy. This is why we have what's known as a negativity bias;[58] that is, we are more averse to losses than we are attracted to gains.[59] Negative stimuli raise our blood pressure and heart rate. We have more negative than positive emotions and have more words in English for painful sensations than for pleasant ones.[60] We respond faster to threats and unpleasant events than to positive opportunities and pleasurable situations. It is better to avoid perceived danger quickly and sometimes be wrong than delay and just once be seriously wrong! These and many other physiological and behavioral human traits are traces from deep evolutionary time, echoes of the lives of our ancestors as they pursued food and sought to avoid being food for other animals. Their struggles have left a negative imprint in our brains.

Psychologists, particularly in Sweden and Norway, have conducted imaginative experiments that show how fears and phobias are acquired and maintained. Most of these experiments have used an approach pioneered by Swedish psychologist Arne Öhman. Investigators first evoke a defensive response by showing either fearful stimuli (snakes or spiders) or neutral stimuli (geometric figures), paired with an electric shock intended to mimic a bite. Then they repeatedly present the same stimuli, but without an electric shock and measure the rate of decline of aversive responses to fear-relevant and neutral stimuli. Aversive responses are usually acquired more quickly to fear-relevant than to fear-neutral stimuli; responses to snakes and spiders always persist longer when they are no longer reinforced than responses to neutral stimuli.

To test whether these results could be due to cultural reinforcement, investigators compared responses to snakes and spiders with far more dangerous, strongly culturally conditioned, modern stimuli, such as handguns and frayed electrical wires. Aversive responses to these modern dangerous stimuli extinguish more quickly than responses to snakes and spiders.[61] A person may acquire an aversive response to fearful natural stimuli merely by being told that shock will be given. Aversive responses to neutral natural stimuli cannot be elicited this way. We may become fearful merely by watching an actor who pretends to react fearfully to slides of fearful stimuli (snakes, spiders, rats) or neutral natural stimuli (berries). But people acquire more persistent aversive reactions when they watch the person's reactions to fearful than to neutral stimuli.[62] Rhesus monkeys respond much as we do to fearful stimuli (toy snakes and crocodiles) and to neutral stimuli (toy rabbits).[63]

Even more striking are the results of "backward masking" experiments in which the investigator shows a slide for only fifteen to thirty milliseconds before it is "masked" by a different slide. Subjects are not consciously aware of having seen the stimulus slide, but slides that contain snakes or spiders elicit strong aversive reactions in most people.[64]

These results show that aversive responses occur even when we are unaware that we have seen the threatening stimulus. We do not respond that way to neutral or fear-irrelevant stimuli. Cultural/learned hypotheses cannot explain these striking results, but they make adaptive sense. Studies of twins show that phobias have a genetic basis. Twins are more like one another than other siblings in their responses to traumatic events, such as dog bites, and their fear of open spaces.[65]

Our ancestors would have benefited from being especially aware and cautious around dangerous situations and objects. Some of those events and objects are still sources of danger in today's world, but many of them are not. The modern world is full of dangers our ancestors could not have imagined. In the last chapter of this book we will revisit the challenges we face when trying to adjust our savanna-adapted minds to life in modern technological societies.

6

Settling Down and Settling In

Vitaly Komar and Alexander Melamid, two dissident Russian artists who defected to the United States, asked a marketing research firm to conduct a survey to assess the attitudes and preferences of adult Americans about the visual arts. Pollsters asked about one thousand people questions such as,

> If you had to name one color as your favorite color—the color you would like to stand out in a painting you would consider buying for your home, for example—which color would it be?

> When you select pictures, photographs, or other pieces of art for your home, do you find you lean more toward modern or more toward traditional styles?

and

> Many people find that a lot of the paintings they like have similar features or subjects. Take animals for example. On the whole, would you say that you prefer seeing paintings of wild animals, like lions, giraffes, or deer, or that you prefer paintings of domestic animals?

The pollsters asked respondents whether they preferred natural or portrait settings, indoor or outdoor scenes, geometric or random patterns, flat or richly textured surfaces, and sharp angles or soft curves.

Komar and Melamid took the findings and incorporated the most preferred features into a single landscape painting, *America's Most Wanted* (figure 6.1). The savanna-like landscape features water (a placid lake), gently sloping hills in the background, and a wooded cliff on the left. The most prominent tree would be easy to climb in the event of sudden

Figure 6.1. Vitaly Komar and Alexander Melamid, *America's Most Wanted*, 1994, oil and acrylic on canvas. Photo by D. James Dee.

danger. Two deer in the shallows of the lake suggest a good source of animal food. The scene features three young people in their reproductive prime plus an older man—George Washington.[1]

Komar and Melamid got similar results from comparable surveys in nine other countries: Russia, Ukraine, France, Finland, Denmark, Iceland, Turkey, Kenya, and China. They also painted the "most wanted" scenes inspired by surveys conducted in those countries. The nine additional paintings had striking similarities to *America's Most Wanted*. All the *Most Wanted* paintings showed savannas with water, easily climbed trees, happy people, and large mammals. Overwhelmingly, people preferred smoothly painted outdoor scenes that looked "real," with blended colors. They liked both wild and domestic animals and human figures, especially children and women, in casual poses. Familiar historical figures had positive valences. Interestingly, as pointed out by art theorist Ellen Dissanayake, neither the artists nor anyone connected with the survey appear to have known about research on environmental aesthetics.[2]

Respondents were asked about paintings they would like to hang in their homes, ones that they would enjoy looking at over and over again.

They may have unconsciously imagined themselves settling in the pictured environments. They were making the evaluations appropriate to the explore stage of our framework.

Jane Austen was not an evolutionary biologist but, as shown by this passage from her 1811 novel *Sense and Sensibility*, she clearly understood how people respond to environments in which they are about to live.

> The first part of their journey was performed in too melancholy a disposition to be otherwise than tedious and unpleasant. But as they drew towards the end of it, their interest in the appearance of a country which they were to inhabit overcame their dejection, and a view of Barton Valley as they entered it gave them cheerfulness. It was a pleasant, fertile spot, well wooded, and rich in pasture. After winding along it for more than a mile, they reached their own house. . . .
>
> The situation of the house was good. High hills rose immediately behind, and at no great distance on each side; some of which were open downs, the others cultivated and woody. The village of Barton was chiefly on one of these hills, and formed a pleasant view from the cottage windows. The prospect in front was more extensive; it commanded the whole of the valley, and reached into the country beyond. The hills which surrounded the cottage terminated the valley in that direction; under another name, and in another course, it branched out again between the two of the steepest of them.

The Dashwood women had lost their estate through inheritance and were forced to move. They had to accept the charity of a relative who offered them a cottage and adjust to a lower social and economic status. The women were not like tourists in a European city, who enjoy exploring its cobbled streets, nooks and corners, compact housing, and quaint shops, but rarely think about what it would be like to actually live there. The women thought about the landscape as a place to live.

As we discussed in chapter 4, during the second stage an individual explores an area and gathers more information about its resources. Exploration may last many days. The environmental features likely to command our attention during the explore stage are primarily those that signal the resource potential of the area and how safe it would be to explore it.[3]

Komar and Melamid used surveys to determine, and paintings to express, the landscape preferences of people in various modern societies. I will also use paintings in this chapter because they (as well as photographs) can be analyzed in several ways. Each can tell us something different about how we evaluate environments as places to live in and how and why we modify them. One of the chief ways paintings reveal our landscape preferences is in the manner that vegetation is depicted. Humans manipulate vegetation for many purposes, often for utilitarian ones, such as clearing forests to make way for settlements or roads, converting wild land to agriculture, or maintaining a surface suited for various sporting pursuits. These utilitarian environments may have aesthetic appeal, but their design is not driven by aesthetics. Songwriters may wax eloquent about waving fields of grain, but we derive aesthetic pleasures in agricultural landscapes from mixtures of fields, pastures, orchards, and woodlots, not from viewing particular crops.

Other modifications to vegetation are designed to please the eye, rouse feelings of patriotism, or provide an intimate space in which to meditate or grieve. Parks, gardens, and cemeteries should reflect our evolutionarily based preferences better than utilitarian environments because we design them to be pleasant places to spend time in. Their design features should entice exploration and settling. Differences in how those places were conceived can tell us much about our preferences for living places. We cannot enter a landscape in a photo or painting, but we do spend time in parks and gardens.

In the paragraphs that follow, I assume that landscape designers modified landscapes to win and keep clients and artists wanted to sell their paintings. Municipal leaders probably also want parks, green spaces, and public gardens to enhance the aesthetic appeal of their communities.

Imagine that we are in Central Park in New York on a sunny afternoon. The park is full of people walking, jogging, bird-watching, skating, sailing model boats, and relaxing on benches and blankets. The park, designed by Frederick Law Olmsted and Calvert Vaux, appears natural, but it is almost entirely landscaped. Among its many important features are several natural-looking but artificially created lakes and ponds, extensive walking tracks, bridle paths, two ice-skating rinks, the Central Park Zoo, the Central Park Conservatory Garden, a wildlife sanctuary, a large area of natural woods, an outdoor amphitheater, seven major lawns, and many small grassy areas. It has something for everybody. No wonder it attracts large crowds of people.

Now imagine that some of our remote ancestors are transported to Central Park. How might they respond to the landscape? To help us think about their reactions, we will use two evolutionary theories—the savanna hypothesis and prospect-refuge theory—that were developed specifically to generate predictions about our responses to landscapes. The savanna hypothesis, which I developed in 1978, asserts that we should be especially attracted to plants that grew in high-quality African savannas because, if they had responded positively to them, our ancestors would have settled in places in which they had good survival and reproductive success.

Jay Appleton's prospect-refuge theory proposes that we decide how to evaluate an unfamiliar landscape by looking for ways to explore it safely. The theory suggests that we should choose routes that would allow us to gain the most information about the environment while exposing ourselves to the least risk. It also predicts that if artists wish their landscapes to be attractive, they should depict a safe route to a refuge that would offer good views (figure 6.2).

So, our ancestors probably would intuitively assess Central Park by paying attention to tree shapes, landscape patterns, and signs of water. If they initially decided that the park merited exploration, they would then decide how to move safely through it to learn enough to decide whether to settle there. In other words, our ancestors would probably respond as we do if they were to suddenly find themselves in Central Park.

We will not use Central Park to test these predictions, but other parks and gardens will serve as models. The savanna hypothesis assumes that our preferences for landscape patterns and tree shapes have been relatively unaltered during the relatively brief time that we have lived in temperate, boreal, and Arctic environments. The trees that dominate resource-rich savannas are typically broader than they are tall, have canopies wider than they are deep, have small, compound leaves, and have trunks that are short relative to their total height (figure 6.3a). Trees that grow along rivers flowing through savannas are taller, narrower, and have trunks that seldom split close to the ground (figure 6.3b). In drier, less productive habitats, trees of many species are shorter; many

Figure 6.2. Ferdinand Bellerman (1814–89), *Road to La Guiara, Venezuela*, oil on cardboard, 1844.

have multiple trunks (figure 6.3c). Our aesthetic responses should reflect both the resources a tree offers (food, shade, safety) and the information it provides about the quality of the environment.

We will test predictions from the savanna hypothesis to find out which tree shapes are most attractive to us. If our aesthetic preferences reflect our savanna origins, we should find trees similar to those that dominate resource-rich savannas especially attractive. Horticulturalists

Figure 6.3. Trees and shrubs of East African savannas. Kenya. (*a*) *Acacia tortilis*, a dominant tree in high quality savannas; (*b*) taller acacias near a stream; (*c*) multi-stemmed shrubs of dry savannas, browsed by gerenuks (*Litocranius walleri*).

should favor and cultivate those trees, and they should make other trees and shrubs more savanna-like by pruning them and selecting mutants with savanna-like shapes.

Similarly, parks and gardens should be designed to resemble African savanna vegetation. All over the world, in parks large and small, we find this to be so. Parks feature scattered trees and shrubs with a grassy understory. Great pains are taken to create water features or the illusion of water, through fountains, ponds, and reflecting pools, and to enhance the quality and quantity of existing water resources for drinking and recreation.

We will test predictions about our responses to unfamiliar environments by looking at gardens that we view from a particular place and by analyzing the structure of landscape paintings. We will also test predictions about our responses during the explore stage of habitat selection by studying the structure of gardens we enjoy by walking through them. But before we test these predictions, let's explore a bit of history.

Tree shapes are clearly drawn in the very early landscape paintings. The Miniature Fresco on Santorini, dating from about 1700 BC, shows a mountain landscape with scattered trees. We cannot identify the species, but they have the spreading form of savanna trees. A fresco showing the king lancing a lion in a mountainous savanna adorns the tomb of Philip II of Macedon, who died in 336 BC. Here we can identify the trees—beeches. They are huge, with spreading crowns, quite unlike the towering beeches of the Macedonian forests that would have been familiar to the artists.[4]

Medieval gardens, the first ones about which we have specific information, were of two main kinds—herb gardens designed for growing food plants and places set aside for recreation, known as "orchards" or "pleasaunces."[5] Most pleasure gardens were surrounded by a wall and had sheltered alcoves where sunshine could enter during winter. They also had arbors shaded by fruit trees, vines, and climbing ornamental shrubs. Many also had artificial hills topped with seats, pavilions, or other structures. Towers, offering even more extensive views, were often erected on mounds or natural hills.

Although aesthetic considerations drove why and how people designed "pleasaunces," other factors also mattered. Modifying large expanses of land is expensive. Large parks can be built only by wealthy landowners or by governments. The structure of landscape gardens also reflects the level of security.[6] Large parks and gardens were created primarily during times when enemies were unlikely to attack. The species of plants that were available, the societal ideals and symbols that plants or structures represented, and the attitudes that society had toward nature also influenced the forms of gardens and parks.[7]

Japanese Gardens Are Savanna-Like

Japanese gardens are especially useful for testing the savanna hypothesis because Japanese gardeners prune many woody plants to alter their shapes. They have also selected and used genetically modified forms of many woody plants. From the eighth through the thirteenth centuries, gardens adorned palaces and villas of royal families and wealthy nobles. They were open and filled with flowers and blossoming trees and contained streams that emptied into ponds and lakes, some of which were large enough to accommodate small boats. Those gardens were designed for the pleasures of the few who could afford them.

No early Japanese gardens survive, but Lady Murasaki Shikibu provided a vivid picture of gardens of the Heian period (794–1185) in her great novel *The Tale of Genji*.

> Genji effected great improvement in the appearance of the grounds by a judicious handling of knoll and lake, for, though such features were already there in abundance, he found it necessary here to cut away a slope, there to dam a stream, that each occupant of the various quarters might look out of her windows upon such a prospect as pleased her best. To the southeast he raised the level of the ground and on this bank planted a profusion of early flowering trees. At the foot of this slope the lake curved with especial beauty, and in the foreground, just beneath the windows, he planted borders of cinquefoil, of red-plum, cherry, wisteria, kerria, rock-azalea, and other such plants as are at their best in springtime. . . .
>
> Akikonomu's garden was full of such trees as in autumn-time turn to the deepest hue. The stream above the waterfall was cleared out and deepened to a considerable distance; and that the noise of the cascade might carry further, he set great boulders in mid-stream, against which the current crashed and broke.[8]

Zen Buddhism's philosophy of simplicity and meditation subsequently influenced all Japanese art forms, but gardens remained savanna-like. They became more sober and changed less seasonally. Evergreen plants dominated them. Rocks and gravel became major construction elements; shrubs were often pruned to resemble rocks (figure 6.4). Small gardens became popular as population density increased and land became more expensive.

According to the savanna hypothesis, trees that are commonly planted in gardens should be broader relative to their height, have relatively shorter trunks, and have small, more deeply divided leaves than rarely planted ones do. Japanese maples (*Acer*) and oaks (*Quercus*), two genera of trees commonly planted in Japanese gardens, are good for testing these predictions.

Three of the twenty-two maple species native to Japan—*Acer sieboldianum*, *A. japonicum*, and *A. palmatum*—dominate those planted in gardens. They naturally grow sideways to seek patches of brighter light in the forest understory. Japanese gardeners do not prune them; they already have savanna-like shapes. The wild *A. palmatum* individuals my wife Betty and I measured do not differ from those in gardens in number of main branches or total height relative to canopy width. Some rarely planted species also have light-seeking growth forms, but most rarely planted species are taller relative to their breadth than the commonly planted ones. They do not resemble tropical savanna trees.

Figure 6.4. Azaleas pruned to resemble rocks, Sento Gosho, Kyoto, Japan.

Although not compound, the leaves of commonly planted maples have five to seven lobes that extend to the middle of the blade, giving them a compound appearance. The leaves of nonplanted species are less deeply divided; eleven have only shallowly lobed leaves, three have lobes that penetrate to the middle of the blade, and two have compound leaves with un-lobed leaflets.

All fourteen species of Japanese oaks (*Quercus*) have simple leaves. Eight species are evergreen; six are deciduous. Oak species commonly planted in gardens do not differ from nonplanted oaks in the number of main branches from the trunk, in height relative to breadth or in relative height of their trunks. All commonly planted species are evergreen with small leaves. In contrast, most deciduous oak species have large, shallowly lobed leaves. Unlike maples, oaks are heavily pruned so that they stay small and have more branches from the trunk than wild-grown individuals.

The natural growth forms of commonly planted conifers in Japanese gardens are not savanna-like, but Japanese red pines (*Pinus densiflora*) that grow in windswept locations, such as seashores and mountain ridges, resemble tropical savanna trees. One of these places, the rocky coast of Omishima Island, is one of the most celebrated scenes in Japan. The red pines' flexible growth form may have attracted Japanese gardeners. Red pines in gardens are pruned so that they are broader than they are tall and have trunks that divide close to the ground. They are trimmed to produce distinct canopy layering much like that of African savanna trees (figure 6.5).

Most conifers are large trees suitable only for parks and large gardens rather than private houses, but many dwarf varieties have been cultivated and planted in small gardens and pots, and made into bonsai. Of fifty-eight conifer species with named cultivars, forty-seven have dwarf and semidwarf varieties, totaling more than two hundred named cultivars.[9] Mutants of twenty-four species with columnar shapes or weeping branches and twigs are commonly planted in parks and gardens worldwide. Gardeners might have sought mutants with acacia-like growth forms, but they are extremely rare.

We can use cultivars of *Acer palmatum*, the maple most extensively planted in Japanese gardens, to test the prediction that they should have smaller and more deeply lobed leaves than their wild ancestors. Wild type *A. palmatum* has green leaves with seven moderately deep lobes.

Figure 6.5. Japanese red pines (*Pinus densiflora*) pruned to resemble savanna trees with short trunks and spreading branches, Kyoto, Japan.

Vertrees divides cultivars into groups called "palmate," "dissectum," "deeply divided," "linearlobium," and "dwarf."[10] For cultivars in each group, I recorded the number of lobes and their depth. I scored depth of lobes on a scale of 1 to 5, where 1 = very shallowly lobed, 3 = lobed approximately halfway to the base of the leaf, and 5 = lobed to the base so that the leaf is effectively compound. Depth of lobes of "wild type" *A. palmatum* varies, but most leaves score between 2 and 3.

The leaves of mutants and wild type trees have the same number of lobes, except in the "dissectum" group whose lobes are so deeply dissected that the leaves are doubly compound. However, depth of lobes has been increased. The leaves of very few cultivars are as shallowly lobed as wild type *A. palmatum*. More than half are lobed to the base, so that they are compound. (figure 6.6)

Weeping or so-called prostrate forms of *A. palmatum* are widely planted. All dwarf cultivars grow no taller than a meter. Many taller forms have multiple stems, are broader than tall when mature, and have drooping branches. Vertrees gives insufficient information for me to assess the sizes of leaves of most cultivars, but the leaves of dwarf plants are much smaller than those of wild *A. palmatum*.

Figure 6.6. Leaves of cultivars of Japanese maples. Many of the cultivars are so deeply lobed that they are actually compound.

In general, these changes accord with predictions from the savanna hypothesis. Compared with wild plants, widely planted mutants have relatively shorter trunks and smaller, more deeply divided leaves. The mutant forms of species that are regularly planted in gardens are small trees with spreading crowns and small compound-like leaves, like the dominant trees of rich African savannas.

Savanna Trees Offer Multiple Benefits

In addition to indicating resource richness, savanna trees also offer safety.

The closer to the ground a tree's trunk divides the easier it is to climb. A dense tree provides good shade, but visibility from its canopy is poor. To test if those traits influence attractiveness, Judi Heerwagen and I recorded people's responses to photographs of *Acacia tortilis*, a common species in resource-rich East African savannas; it often dominates safari advertisements. We selected photographs of individual trees that varied in height-to-width ratio, height at which the trunk divided, and extent of canopy layering. We grouped them so that one category (trunk height, canopy density, canopy layering) varied but the others did not (figure 6.7).

We asked participants to rate each tree on a continuous scale: 1 to 2 = "unattractive," 3 to 4 = "moderately attractive," and 5 to 6 = "very attractive." Participants could score a tree as, say, 3.6 rather than 3 or 4. We recorded ratings of seventy-two people entering or leaving the University of Washington Bookstore and thirty in a restaurant on the University of Washington campus. They ranged in age from eighteen to sixty. We approached people and asked them if they would mind filling out a questionnaire on the role of trees in environmental aesthetics. Instructions on the survey read, "In the attached photo-questionnaire, we ask you to rate the relative attractiveness of a number of trees. There are six trees on each page, with the rating scale under each. Circle the point on the scale which best corresponds to your opinion of the tree's attractiveness." They were asked to rate the trees on each page before they continued to the next and not to turn back once they had completed a page.

We tested four hypotheses: (1) Trees whose trunks divide closer to the ground should be more attractive than trees whose trunks divide higher

Figure 6.7. One of the pages of photos of *Acacia tortilis* trees that Judi Heerwagen and I used to test people's aesthetic responses to tree shapes.

up. (2) Trees with moderate canopy density should be more attractive than trees with low or high canopy density. (3) Trees with a high canopy layering should be more attractive than trees with low or moderate layering. (4) The broader a tree's canopy relative to its height, the more attractive it should be.

As we predicted, low trunk height, canopy layering, and wide canopies positively influenced people's scores.[11] Contrary to our predictions, however, the canopy-width-to-canopy-height ratio did not influence a tree's attractiveness. We were not able to assess the impact of canopy density because canopies appeared more similar to one another in the black-and-white photo-questionnaire than in the original color photos.

Ghosts in Western Gardens

Western gardens are also rich sources of information for evaluating the savanna hypothesis. Western gardening began in the Middle East, a dry region with brutally hot summers. Agriculture was restricted to valley bottoms and depended on spring meltwater from mountain snow. Water was scarce and valuable. In striking contrast to the dry natural landscapes that surrounded them, Persian gardens were walled oases of fruit trees and water. They were designed as places to sit and escape from the sun's heat. Persian gardens were models for Islamic gardens, which spread eastward to India and westward to Turkey, North Africa, and Spain.

We have only indirect evidence of the forms of those ancient Persian gardens. One source is an immense carpet of silk, gold, silver, and jewels, known as the Spring of Khosrow Carpet, which apparently represented the spring garden of King Khusrau I (531–579). Arabs discovered it when they conquered the Mesopotamian city of Ctesiphon in 637. They reported it to be about 450 feet long and 90 feet wide. Unfortunately, the carpet was cut up and the fragments handed out to the troops as the spoils of war. Gardens portrayed on surviving Persian carpets have a small border of flowers, followed by a wider one of trees. The main part of the garden is divided into four sections, separated by canals with fish. Each of the four sections is divided into six squares where flowerbeds alternate with beds containing plane and cypress trees.

Marco Polo traversed the Near East during the 1260s. He described a garden "planted with all the finest fruits in the world" and with "four conduits, one flowing with wine, one with milk, one with honey, and one with water." We must take his description, especially of the conduits, with many grains of salt, but it does suggest that real gardens were similar to what is portrayed on Persian garden carpets that survive today. Our remote ancestors would have felt comfortable in them.

No ancient Persian gardens remain, but a few Moorish gardens survive in southern Spain. Moors invaded Spain in 710 and ruled much of the country for centuries. Spaniards did not reconquer the Kingdom of Granada, the last bit of the Moorish empire, until 1492. Despite strenuous efforts to eradicate Islamic belief and Arabic learning, some aspects of Arab culture survived; as we will see, gardening is one of them.

Walking in Large Gardens—Prospects and Refuges

Jay Appleton's prospect-refuge theory predicts that designers should provide garden visitors with a succession of refuges, each one offering a view of a different part of the garden. Views should feature areas that offer water, the promise of resources, and safety. Designers may ensure that visitors see these views by directing walkers along a specific route of travel indicated by clearly marked rock or gravel paths.

Japanese gardens fit these predictions. Paths curve; rocks, trees, or shrubs frequently hide their courses.[12] A walker emerges from one refuge with a new prospect view. After a short walk through an open area, another refuge (a group of trees and shrubs that obscures other parts of the garden) appears. Multiple prospect views also foster an impression that a garden is much larger than it actually is.

Western gardens, although they superficially look very different, share some structural features. For example, the Great Avenue at Castle Howard, Yorkshire, England, is about five miles long, most of which is straight, but it has a series of little hills each of which reveals a new vista. Architectural features—an early eighteenth-century obelisk, a Victorian column, and a number of gateways also generate variety.

The most elaborate formal garden designed around a specified route of travel also is the largest—the gardens of the Palace of Versailles. The garden is so large, at eight hundred hectares, that King Louis XIV actually wrote an itinerary for viewing his garden. Entitled *La Maniere de Montrer les Jardines de Versailles*, it not only indicates the route, but also commands the visitor to look at views in a specific way. "Leaving the Chateau . . . go on to the terrace. You must stop at the top of the steps . . ." The gardens at Versailles are so extensive that it would have taken most of a day to complete the route, even by carriage. Following Louis's itinerary, a visitor would move through the garden from one fountain to another, never losing sight of water. Sited on flat terrain, Versailles was regularly short of water. Indeed, the specified itinerary allowed fountains to be turned on in a particular order. A "secret" route ran parallel to the king's. Boys ran along it, signaling with flags when it was time to turn fountains on and off.

Italian Renaissance gardens had ample supplies of naturally flowing water. Designers manipulated views to generate surprises. Leon Battista Alberti (1404–72), who wrote extensively about principles of garden design, suggested "I would have it stand pretty high, but upon so easy an Ascent, that it should hardly be perceptible to those that go to it, till they find themselves at the Top, and a large Prospect opens itself to their view."[13]

Many little-modified Italian Renaissance gardens remain. The most perfectly maintained one is Villa Lante, Bagnaia, begun in 1564 and extensively restored in 1954. The long, slim rectangular garden descends in terraces down a wooded slope. Water flows down its central axis from a grotto at the top to a square pool at the bottom (figure 6.8). Villa D'Este, begun around 1550, has a similar design but with the addition of elaborate fountains. A visitor is encouraged to enter at the bottom and follow gently rising paths. At the top a visitor is rewarded with a magnificent view of the city of Tivoli. A similar design with a stairway thickly shaded with trees and shrubs, from which a walker emerges to a spectacular view of Rome, is found in Villa Medici.

England's expansive landscape gardens can be traversed via multiple routes, but Lancelot "Capability" Brown (1716–83) designed his gardens so that travelers were repeatedly surprised. He planted strips of woodland to conceal undesirable landscape features. He used clumps and isolated groups of trees to frame distant views. Brown also created many lakes by damming streams. He concealed dams so they were hidden until a person got very close to them (figure 6.9).

Moorish gardens also were designed for walking. The best surviving ones, the Generalife and the Alhambra, are in Granada, Spain. The

Figure 6.8. Lower Garden, Ville Lante, Bagnaia, Viterbo, Italy.

Figure 6.9. Burghley House, built 1565–87 by William Cecil, Lord Treasurer to Queen Elizabeth I; lake designed by Capability Brown, 1775–80; Stamford, Lincolnshire.

Alhambra, built around 1238, is an *alcazaba* (military fortress), an *alcázar* (palace), and a *medina* (small city) all in one. The Generalife is a complex of gardens built over many centuries. In 1958 a fire destroyed part of it. Restoration workers found that the original garden beds were well below the paths; the tops of flowers would have been roughly level with the paths. The Court of the Lions is the purest surviving Moorish garden in Spain. Water rises in adjacent rooms and fountains, runs along narrow channels past slender pillars that line the garden, and forms four rivers that divide the beds. Today's beds are level with the paths; the original ones were probably lower.

Although the general layouts of Granada's gardens are probably very similar to the original Moorish design, the fountains are recent Spanish additions. Cypress trees, which had religious significance in Persia, were part of the original Moorish plantings, but many plants, which include salvias and yuccas from the Americas, are recent introductions. We do not know if or how the Moors manipulated tree forms. Currently, most trees in the Generalife and the Alhambra are heavily pruned, a necessity in such small spaces, but few of them are modified to have savanna-like forms. Other than oranges, none of the trees or shrubs bears edible fruits.

Distant Views Are Important

As we saw in chapter 4, people sitting at the entrance to Abauntz Cave in Spain had expansive views of plains and mountains. With that view they could see distant prey and spot approaching people. Long ago gardeners learned how to use expansive views of areas beyond a garden's borders to give an impression that it is much larger than it really is. Such borrowed scenery, as it is called, also enhances the perceived power and authority of the garden's owner (figure 6.10). Borrowed scenery is especially important for gardens designed to be viewed from a nearby spot. We consider this in the next section.

Assessing Prospects from Afar

When we encounter an unfamiliar landscape we make a rapid unconscious assessment to determine whether to stay and explore it or move on.

Figure 6.10. Borrowed scenery, Bodnant Garden, North Wales.
The garden ends at the low wall just behind the pond.

A designer of a garden that is to be viewed from a nearby spot confronts the same limitations as painters and photographers. Yet, a gardener has a major advantage. A painting or photo, once completed, is forever the same, although today's digital photos are easily modified. A garden changes with weather, time of day, and season. Trees burst into leaf, turn colors in autumn, and drop their leaves. Flowers bloom and fade. A succession of flowers, enhanced by ripening fruits, generates seasonal color changes. Just as one cannot wade into the same river twice, one cannot view the same garden twice. A person can experience something new with each visit.

Many small Japanese gardens are viewed from a particular spot. The main attraction is often water, real or simulated. Japanese garden designers are experts in creating an impression that these small gardens are larger than they really are. They sometimes build boats half the normal size so that the ponds in which they float seem to be large. They employ horizontal lines, such as a stretch of sand or gravel raked into flowing lines, to create a sense of spaciousness. They magnify apparent depth by placing large rocks and plants near the viewpoint, smaller ones farther away.

Japanese designers also create a sense of distance by placing plants and materials of brighter colors near the viewing spot and darker shades in the distance, and by planting trees and shrubs with larger leaves nearby, woody plants with smaller leaves farther away. A curving path that winds among trees and shrubs appears to be far away. A sense of greater size is sometimes created in small gardens by making a plot of lawn or sand slightly narrower farther from the viewing point, a device that makes the garden appear to recede more rapidly.

Garden designers in the West also took advantage of borrowed views. Cattle and sheep roamed large gardens. They kept the lawn closely clipped and provided owners with dairy products and meat. Walls, hedges, or fences controlled livestock movements and provided privacy. These English gardens were inward looking and visually tied to the house. In 1738, a simple but important innovation—the ha-ha—dramatically changed appearances. A ha-ha is a ditch dug around those parts of a garden to be visually joined to the landscape. It was wide and deep and had one steep side, so that cattle could not cross it in either direction. Ha-has were designed to be invisible from the house to create an illusion that the garden and surrounding countryside were united. The name "ha-ha" came to be used because people were surprised when they suddenly encountered the unexpected barrier.

Landscape gardeners were much influenced by landscape painters, particularly Claude Lorrain (1600–1682) and Gaspard Dughet (1615–75). They both painted many landscapes near Tivoli, about fifteen miles east of Rome. That area is hilly; the River Anio plunges down its valley in a series of cascades. The ruin of the temple of the Vesta is perched on a jutting corner of a hill near Tivoli. Dating from the first century BC, the temple of the Vesta appears in more than forty of Claude's surviving paintings and in a high proportion of Gaspard's. Garden designer William Kent (1685–1748), first trained as a painter, studied in Italy from 1712 to 1719. He introduced imitations of the temple to English gardens; it became the most frequently imitated ancient building in British landscapes. More than twenty were built; even more were built on the continent. Kent viewed gardens as landscapes through which a visitor could proceed from one "landscape picture" to another. Each "picture" would have an association with nature and some images of the past in the form of urns, statues, temples, and ruins.

Changes recommended by landscape architects to their prospective clients can help us understand what motivated them. Humphrey Repton (1752–1818), a leading British landscape architect, provides the most complete record. He prepared "before" and "after" drawings for his clients, bound his drawings in red covers, and explained in an accompanying text why he proposed those changes. Judi Heerwagen and I analyzed eighteen "before" and "after" drawings in Repton's Redbooks to test whether the changes he recommended fit predictions from the savanna hypothesis and prospect-refuge theory.[14]

As we predicted, Repton often added trees and copses to open fields. He added groups of trees at the water's edge, a feature that protects people when they are drinking, resting, or bathing (figure 6.11). Our savanna ancestors were less alert when engaged in those activities and, therefore, benefited from the protection afforded by trees. Repton also frequently altered straight borders demarking pasture and woods by planting trees in open spaces to create uneven edges. He also added scattered copses of trees, removed trees to open up distant views, and opened woods to allow both physical and visual access. He modified trees so that their trunks split close to the ground, making them easy to climb, and added flowers and shrubs. He removed trees in about half of the landscapes to open views to the horizon. In his book *The Art of Landscape Gardening*, Repton provided the rationale for the changes he proposed. He noted that too many trees "make a place appear gloomy and damp."[15]

Water features are the most striking part of Repton's designs. He frequently added bodies of water and enlarged existing ones to make them more conspicuous. In about half his designs, he enhanced the view of water features. He added rocks to a brook to give it "a rippling, lively effect which is highly preferable to a narrow stagnant creek." Repton believed that "it is only by such deceptions that art can imitate the most pleasing works of nature."[16]

Repton also added cattle, sheep, and deer, and where appropriate, boats. He defended doing so by saying that "both are real objects of improvement, and give animation to the scene." His addition of animals is especially interesting because doing so went well beyond the common scope of a landscape architect's commission.

Two technical inventions—the lawn mower and the greenhouse—exerted an important influence on Western garden design. Prior to the

Figure 6.11. *Above,* Sketch showing the "before" view of the water at Wentworth, Yorkshire. *Below,* Lifting the flap reveals the "after" view with Humphrey Repton's proposed improvements.

invention of the lawn mower by Edwin Budding in 1830, closely cut lawns were maintained either by sheep and cattle or laboriously with a scythe. A small landowner who lacked sheep or a team of gardeners could afford a lawn mower. By the 1870s the lawn mower had triumphed; scythes were rusting in tool sheds.

The greenhouse was invented in 1829 by Nathaniel Bagshaw Ward, a doctor, an amateur entomologist, and an ardent grower of ferns. He got his idea when a fern appeared in a glass bottle in which he had buried the chrysalis of a sphinx moth. He concluded that plants would prosper in a bottle where the air was free from soot. He placed the bottle outside the window of his study and was delighted that the fern and other plants continued to grow. Stimulated by this simple experiment he built a glass-enclosed case about five feet high with a perforated pipe around the top through which water rained on the plants.

It did not take long for greenhouses to be widely used. With a greenhouse a gardener could grow large numbers of tender annual plants to transplant at specific times. When they had finished blooming, plants could be dug up and replaced with another batch of annuals from the greenhouse. Greenhouses for the first time made it easy to maintain a garden with dramatic seasonal changes in flowering herbaceous plants.

Woody plants are also sources of seasonal color variations. Autumn leaves of deciduous trees are an obvious example, but many cultivars have brightly colored or variegated leaves during the summer growing season. Maple mutants with bright red petioles and red samaras (fruits) are also extensively planted, as we will see in chapter 8.

"A Beauteous City for the Dead"

The term "cemetery" is derived from a Greek word that means "sleeping place." Greeks moved cemeteries outside the city for sanitary reasons, but their main motivation was to provide a place for visitors—the dead being unable to appreciate their surroundings—to establish contacts with nature and to achieve a sense of the continuity of life. Few objects were placed in Greek graves, but elaborate and brightly painted stelae and statues often marked the graves to ensure that the dead would not be forgotten.

The Greek concept of cemetery settings was adopted in America's most important cemetery, Mount Auburn, which was established in Cambridge, Massachusetts, in 1835 and modeled on the Père Lachaise Cemetery in Paris. Foreign dignitaries were regularly taken to the cemetery. It was quipped that Bostonians had only two ways of entertaining important guests—a formal public dinner and a drive to Mount Auburn. Mount Auburn was much admired and was imitated in cemeteries throughout the eastern and midwestern United States.

Lady Emmeline Charlotte Elizabeth Stuart-Wortley, who was taken to both the Harvard campus and Mount Auburn in 1849, wrote an enthusiastic description:

> The finely diversified grounds occupy about one hundred acres, in general profusely adorned with a rich variety of trees, and in some places planted with ornamental shrubbery: there are some tombs graced with charming flower beds. There are also some pretty sheets of water there: it is divided into different avenues and paths, which have various names. Generally they are called after the trees of flowers that abound there, such as lily, poplar, cypress, violet, woodbine, and others. It is, indeed, a beauteous city for the dead. The birds were singing most mellifluously and merrily—it was quite a din of music that they kept up in these solemn but lovely shades. The views from Mount Auburn are fine and extensive. There are some graceful and well-executed monuments within its precincts.[17]

Color Vision and the Preference for Blue and Green

This survey of Western and Japanese gardens shows that they share a number of features. But are these common features the result of universal human aesthetic preferences? They may be. As we saw at the beginning of the chapter, surveys carried out for Komar and Melamid yielded similar results in ten countries. People in all countries preferred similar structures, and they all preferred blue, followed by green.

Why did people prefer blue and green? One possibility is that these two colors have been dominant features of the natural environment for billions of years. They are inevitable by-products of the physical properties of water and of the compounds in plant leaves that absorb light and support photosynthesis. They already existed when color vision evolved several hundred million years ago. At that time very few colors would have been found on Earth. In addition to blue, green, and brown, there would have been the colors of the rainbow, reflected colors of crystals, and the red of spilled blood, but not much more. The most striking colors of nature—the brilliant plumage of birds and the rich colors of flowers and fruits—evolved because their bearers were more easily seen and color could send a signal of vigor, fitness, or ripeness. But those colors were more conspicuous only to an animal with color vision.

My explanation of Komar and Melamid's findings is not universally shared. Arthur Danto, a philosopher and art critic for the *Nation*, asserts that our preferences for pictures are culturally produced. He believes that we prefer the colors our culture makes familiar.[18] He suggests that our landscape preferences have been determined by what is pictured in calendars. He is correct that, in response to the questions about what types of art people have in their homes, 91 percent of Kenyans mentioned prints from calendars, but his explanation fails to explain why those scenes are selected for calendars. Calendar producers surely want images people will enjoy looking at. Their selections must be guided by an intuitive sense of what we find attractive. In my view, Danto has the causation backward.

Responses to Individual Trees

Investigators have used people's responses to computer-generated trees or photos of trees to test whether we find savanna-shaped trees particularly attractive. Robert Sommer and Joshua Summit showed drawings of columnar, globular, fan, broad oval, and narrow conical shaped trees, as well as eucalyptus, oak, conifer, palm, and acacia, to college students in Australia, Brazil, Canada, Israel, Japan, and the United States. They all rated spreading and globular trees higher than conical and columnar ones.[19] College students from Zimbabwe, South Africa, Estonia, Italy, Switzerland, and the US-Mexico border also preferred trees with an acacia-like shape.[20] The rank order of tree shapes was spreading, globe, fan, oval, conical, and columnar. These results strongly support the savanna hypothesis. Participants also scored trees more highly if they were common where they had grown up, indicating that early exposure to a tree increases its attractiveness. In laboratory experiments, people have lower blood pressure when they view trees with spreading rather than rounded canopies.[21]

Children and Trees. All experiments I have described so far were conducted with adults. We expect children to respond differently to trees and their responses should change with age as their ability to climb improves.

Few experiments have been conducted on children's tree-climbing activities, in part because of ethical concerns. However, Coss and Moore tested whether young children with little or no tree-climbing experience understood that trees could be antipredator refuges and sources of shade.[22] Because our female ancestors were better climbers than males during most of human evolution, Coss and Moore predicted that preschool girls would value trees as antipredator refuges more highly than preschool boys did. They tested this prediction by asking children where they would seek refuge from a predator.

In one experiment Coss and Moore used a computer-generated scene with a spreading tree next to a rock outcrop that contained a crevice a child could squeeze into but that was too narrow to admit a lion. Both children and the lion could climb the outcrop. Coss and Moore took snapshots from different vantage points in this virtual world and organized them into a picture book. They used eighteen pictures in a narrated tour of the three potential refuge sites. The narrator pointed to relevant features in each picture before turning the page. Then the lion was presented, and the assistant continued with the next set of pictures that reviewed the tour and reminded the children that they had seen three places where they might hide. Finally, the assistant asked each child, "where would you go to feel safe from the lion?"

As they had predicted, girls selected the acacia tree more often than either the crevice or top of the boulder. The boys, on the other hand, did not favor any one of the three. Combined, boys and girls selected the tree for a refuge more often than they selected the top of the rock. Evidently they understood that a lion could climb the rock but not the tree.

In another experiment, Coss and Moore presented children from three cultures with silhouettes of trees that differed in crown height and width to determine if crown shape influenced choice of tree for aesthetic appreciation, accessibility, visibility, sleeping, and refuge from a lion. They used four trees—an Austrian pine (*Pinus nigra*), an African fever tree (*Acacia xanthophloea*), an unbrowsed *Acacia tortilis*, and a heavily browsed *A. tortilis*. Children could "climb" the trees with their fingers, but they could not tell how large they were. They predicted that children would select a tree with a wide crown rather than a tall crown as the prettiest tree.

Children were first asked if they climbed trees, how often, and where they climbed them. They were then asked four questions, in random order: "Which tree is the prettiest?" "Which tree would you climb to see better?" "Which tree would you choose to sleep in?" "Which tree would you climb to hide?" The final, and most provocative question was, "A wild lion has escaped from the zoo and was seen nearby. Which tree would you climb to feel safe?"

To see better, children preferred the pine, fever tree, and browsed *A. tortilis* over the unbrowsed *A. tortilis*, the tree with the densest canopy. However, most of them preferred the unbrowsed *A. tortilis* as the tree in which to hide, to sleep in, and to feel safe from the lion. This result is not surprising; three- to five-year-old children show a good understanding of how obstructions restrict another person's field of view when they play hiding games.[23]

More than half of the children selected the Austrian pine, a tree with a tall narrow crown, as the prettiest, a result that differs markedly from the aesthetic preferences of adults in most experiments. It also conflicts with predictions from the savanna hypothesis. Apparently, for reasons we do not understand, aesthetic appreciation of tree shapes changes during transition from childhood to adulthood because college students strongly preferred the unbrowsed *A. tortilis*; only 5.4 percent preferred the Austrian pine.

In another experiment, Coss and Moore used the same tree silhouettes to test possible sex differences in tree preferences. Half of three- to four-year-old preschool children at day-care centers in Davis, California, were presented with the four trees in the same positions on the page as in the second experiment. The others were presented with a rotated layout in which the Austrian pine appeared in the lower right-hand corner. They asked three new questions: "Which tree would be the most difficult to climb?" "Which tree would you stay under to keep cool on a hot day?" "Where in that tree would you climb to feel safe?"

Children judged the fever tree to be the most difficult to climb, but not by very much. They strongly preferred the unbrowsed *A. tortilis* as the tree to stay under to keep cool and the tree in which they would feel safest from the lion. Both four-year-old boys and girls selected a significantly higher refuge site in the Austrian pine than did three-year-old boys. Four-year-old girls selected slightly higher refuge sites in the fever tree than did three-year-old children. Four-year-old girls selected safe sites closer to the edge of the canopy of both the unbrowsed and browsed A. *tortilis* tree than did four-year-old boys. Small branches far from the trunk are safer because some predators, especially leopards, can climb trees but cannot crawl out on small branches without breaking then.

These results are unlikely to be due to experience because all children had climbed only commercially built structures in backyards, parks, and school grounds; none of them resembled trees. The children appeared to possess knowledge about the refuge value of trees that they could not have derived from prior experience. Their judgments appear to be adaptive in the circumstances under which our ancestors would have climbed trees. Baboons, langurs, and macaques also choose outer branches for refuge from heavier-bodied predators and preferentially sleep on small, outer branches.[24]

Symbolic Significance of Tree Shapes

Inevitably we attach symbolic significance to trees. We associate them with positive values, such as permanence, stability, trustworthiness, fertility, and generosity.[25] Trees are prominent in children's stories and myths of many cultures. The three parts of a tree—root, trunk, and crown—have been thought to reflect the infernal, earthly, and heavenly domains.[26]

Although most of us are attracted to trees similar in form to those that dominate resource-rich African savannas, we also find trees with other shapes to be beautiful. Trees that are tall and narrow are widely planted near houses and along roads. Tall, slender cypresses are prominent in Middle Eastern and Mediterranean gardens. The Lombardy mutant of a European poplar is common in Western gardens. The savanna hypothesis cannot explain why those trees have strong aesthetic appeal.

A plausible alternative hypothesis is that their appeal derives from our positive responses to sunlight. We require sun exposure to synthesize vitamin D, which we need for development of the musculoskeletal system, preventing rickets and osteoporosis, and preventing chronic diseases such as type 1 diabetes and rheumatoid arthritis. Daylight also regulates the body's twenty-four-hour circadian rhythms. When we are exposed to insufficient daylight or artificial light, melatonin levels increase, causing us to become drowsy and depressed.

The sun arcs across the sky during the day, but sunlight comes primarily from above. Darkness reigns below. Most religions postulate that heaven or its equivalent lies above in the sky. Hell or its equivalent lies in the dark interior of Earth. Spiritual help comes from above. Positive emotions are associated with events above us. We speak of having a "high" or feeling "low." We have "uppers" and "downers." We speak of being "lifted up" or "dragged down." We have higher and lower thoughts. When people are asked to evaluate whether words shown on a screen are positive or negative, they respond more quickly when a positive word is flashed at the top of the screen or a negative word is flashed at the bottom of the screen than the reverse.[27]

Our deep intuitive sense that "happy is up" may explain why we associate striving to achieve "higher states of awareness" and greater status with symbols that point upward. Tall narrow trees may evoke "uplifting" emotional responses, especially on flat terrain. The spires and steeples of religious buildings may also reflect this fundamental emotion. Nations compete to have the highest buildings, not those that occupy the greatest amount of ground space.

The Significance of Water

We can survive several weeks without food, but we must drink daily. Omega-3 amino acids, which we need for building large brains, are difficult to get. Aquatic animals, such as mollusks, are good sources. We are surprisingly agile in water. Infants prior to six months of age exhibit appropriate movements and breath control when placed in water. Water always increases the attractiveness of environments. We highly value water in our initial responses to landscapes (encounter stage). We also value water when we evaluate environments for longer-term use (explore stage). We flock to shores of rivers, lakes, and oceans for vacations. We add water, sometimes at great expense, to our private yards. Landscape architects enhance water in public parks and gardens. We pay more for homes that front on water or have views of water.

Water has probably always been a major feature of gardens, but evidence of water disappears quickly when a garden is abandoned, so we have only indirect evidence about water in most ancient gardens. We know about water in Egyptian gardens because it is pictured in tombs, sculptural relief, and paintings with decorative geometric ponds and canals with fish, lotuses, and papyri.

Mesopotamian engineers created networks of lakes, reservoirs, and canals. They were destroyed, restored by Nebuchadnezzar between 604 and 561 BC, and destroyed again by the Persians. As described by Greek historians Strabo and Diodorus, the Hanging Gardens of Babylon, created by Nebuchadnezzar to please his Persian wife (605 BC), contained conduits and fountains. Engines drew water out of the river Euphrates and supplied it to the garden via concealed conduits.

The word "fountain" initially meant a spring of water. Many Greek springs were dedicated to gods, goddesses, nymphs, and heroes. Water delivered to carved basins was often used for drinking. One Corinthian fountain had a statue of the horse Pegasus with water flowing from its hooves. Another included a bronze statue of Neptune standing on a dolphin, from which water flowed. Gardens and then cities grew up around springs.

Rome is a city of fountains, but the number today is miniscule compared with the 1,212 public fountains and 926 public baths that existed when the Goths sacked the city. Only five of the original fountains survive. These impressive structures were built where they could be fed by water from mountain springs and rivers.

As we have described, traditional Persian gardens contained four essentials: water for irrigation, display, and sound; shade trees for shelter; flowers for scent and color; and music. In much of the Islamic world (India, Pakistan, the Middle East, northern Africa, Spain, and Portugal), where water was scarce, architects made the most of small amounts.

Designers of European Renaissance gardens generally worked with the natural terrain, but Baroque gardeners often built ramps, terraces, and steps, many of which gushed with water. Renaissance and Baroque gardens often had water-driven devices that produced a variety of noises.

Chinese and Japanese garden designers also extensively used water, often in the form of a sinuous pond, interrupted by hills, mounds, rocks, and trees, positioned such that only part of the water can be seen from one place. Japanese designers perfected the art of creating a dry streambed,

Figure 6.12. Rock garden with gravel raked to simulate waves.
Upper garden, Rengejo-in, Koyasan, Kyoto.

first developed by the Chinese. Carefully placed rocks, sand, and gravel and raked patterns create an image of moving water (figure 6.12).

An eleventh-century manual, the Sakuteiki (Memorandum on Garden Making), recommends the following procedure for constructing a stream. "In making a stream in a garden, place the rocks where the water turns: then it will run smoothly. Where the water curves, it strikes against the outer banks, and so a 'turning stone' should be laid here and there as if forgotten. But if too many stones are placed along the stream, while it may appear natural when you are close by, from a distance it will seem as if they had no purpose. Moreover, an excess of rocks will make the course seem one of stone rather than of water. Thus, the water effect will be spoiled."

Water and Real Estate Prices. Realtors have long known that market prices increase if houses have views of water. Data from 6,949 home sales from 1984 through 1993 in Bellingham, Washington, show that water frontage increased the value of a home by 126 percent compared to a comparable nonview, nonfrontage home. An unobstructed view of Bellingham Bay added even more to the market price; a mountain view added much less to its value. The market value of a house decreased slowly with its distance from the bay.[28]

Houses in eastern Massachusetts with a water view sold for more than similar houses without one.[29] So did three thousand houses in the Netherlands built after 1970. The largest increases in prices (28 percent) were for houses with a garden facing water connected to a sizeable lake. Sale price increased 8–10 percent if a house overlooked water.[30] The size of the water body had little influence on sale price.[31] This response suggests that the benefit of the water is for drinking rather than transportation.

Aesthetics of Imaginary Environments—Heaven and Hell

We know that we will die, but throughout recorded history, and presumably long before that, people did not want to accept that death is final. They resolved the psychological pain of this grim prospect by constructing visions of life somewhere else after death on Earth. Prehistoric people easily observed evidence of life after death. Dead plants appeared to spring to life after rains or in spring, amphibians emerged from mud, and migratory birds reappeared from unknown places after long absences.

Perceptions of heaven and hell tell us something about the features we would like to find in the environment of our next life. Limits to human imagination guarantee that concepts of the "otherworld" are strongly influenced by the world we do know. We imagine life in those worlds in a body similar to the one we live in. Heaven is similar to Earth but without most unpleasant parts of the earthly environment. Australian aborigines thought the land beyond the great water in the sky was like Australia but more fertile, well watered, and full of game. For the Comanche, the land where the sun sets is a valley wider and longer than their own valley, where there is no darkness, wind, or rain, and game is abundant.[32]

Because people have thought that life after death continues in familiar ways, the dead were typically buried with possessions they would need in their next life. Rich people were assumed to need their servants in their next life. Tombs of city kings and their consorts, found at Ur in Mesopotamia, dating from before 2000 BC, contain many servants, soldiers, courtiers, and ladies-in-waiting who were killed and buried with them. In China, under the Shang Dynasty (from the eighteenth to the eleventh centuries BC), kings and noblemen were buried with a substantial escort of servants.

Depictions of the two options were clearly designed to reinforce the desirability of getting to heaven but avoiding hell. What are they like?

Heaven

Christian heaven was filled with beautiful trees that provided bountiful fruit and perpetual shade. Jesus compared the Kingdom of God to a great tree in which many birds make their nests. In the fifth century BC, Pindar described the Isles of the Blessed: "The corn grows by itself, the fig needs no grafting, vines are always in flower and olives in shoot, honey drips from the oaks and water splashes down the mountain-sides. The cows and goats do not need tending. The climate is perfect, and the place is totally untouched by the ungentlemanly fingers of trade and commerce." Heaven is populated with the same organisms that support a hunter-gatherer's existence on Earth; pure water is always available.

According to the Koran, believers go to enclosed gardens full of maidens and fountains. "They shall feel neither the scorching heat nor the biting cold. Trees will spread their shade around them, and fruits will hang in clusters over them." A Muslim tradition describes a tree in paradise so large that a person on a horse could ride beneath its shade for a year. In Hindu legend a colossal rose-apple tree that gives shade to the whole Earth grows on Mount Meru. The juice of its fruits, which are as big as elephants, forms the river of immortality.

Hell

Christianity inherited the concept of hell from Judaism and mythologies of the Ancient World. Judgment and punishment of some sort were necessary consequences of Christian doctrine. If God had sent his son into the world to save all who believed in him, those who did not accept him and his teaching would not be saved. It did not necessarily follow that they must be physically tortured after death, but they needed to suffer in some significant way. The God of the Old Testament was a god of thunder and lightning; fire was one of his principal weapons.

Descriptions of hell were vague until Virgil described it in his epic poem, the *Aeneid*. The most detailed description of the Christian hell is found in Dante's *Inferno*, part of his fourteenth-century *Divine Comedy*. In the *Divine Comedy*, Virgil himself is the guide through hell and purgatory. Dante's hell has nine circles or layers that correspond to the nine spheres that characterize the heavens. The deeper down into the pit, the greater is the degree of wickedness and the more terrible the punishment.

In most accounts of hell the principal physical tortures are from fire and heat. A lake of flame descends from the lake of fire and brimstone in the book of Revelation. Torrents of fire and heat also are prominent in the hells of other religions. Hell is described less vividly in the Koran than paradise is, but it is not a pleasant place. In Buddhist and Hindu hells, fire is a major agent of punishment, but punishment for most victims is temporary. Sooner or later, most sinners are released and sent back to lead fresh lives on Earth. Only the exceptionally wicked are condemned to eternal punishment.

Pictorial representations and verbal descriptions of Christian hell portray the landscape as bleak and barren, with gloomy valleys, jagged cliffs, stony plains, whirlwinds of fire, rivers of flame and pitch, fuming pits, putrid swamps, and yawning chasms. Instead of sweet music and songs of praise and thanks to God, hell resounds with screams of agony, wailing and gnashing of teeth, and desperate but useless requests to God for mercy.

Oriental hells also have grim rulers and monsters. The Tibetan Book of the Dead, a guide to the experience of dying, describes the judgment in Yama's court. A painting of the scene shows him wearing a human skin as cloak, a girdle of human heads, and a headdress of skulls. He holds a sword and a mirror in which every human action is reflected. The good and evil actions of the dead are weighed in his presence; the wicked are led off to punishment in the eight hot hells and the eight cold hells.

The hell of the Jōdo-shū or Pore Land sect in Japan also has eight hot divisions and eight cold, presided over by Emma-ō, the Japanese equivalent of Yama. He is shown dressed as a judge with a book in which all human actions are recorded and with two severed heads beside him.

The vivid and detailed accounts tell us more about the ranking of the seriousness of various sins during the fourteenth century than it does about environmental aesthetics. Nonetheless, the themes of water contaminated by pathogens and excrement and lack of resources to support a meaningful existence are common to all descriptions of hell. Heavens are invariably safe places with abundant resources. There are no resource-rich savannas in hell!

As this chapter demonstrates, by employing a variety of perspectives generated by evolutionarily based theories, we can now offer plausible explanations for the design of gardens, our aesthetic responses to tree shapes, and how we imagine the otherworld to be. More hypotheses and more tests of them lie in the future.

7

A Ransom

in Pepper

The first recorded plant-hunting mission, dispatched to "the land of Punt" on behalf of Queen Hatshepsut of Egypt in 1495 BC, sought spices. The expedition, led by Prince Nehasis, sailed up the Nile, crossed over to the Red Sea, and then headed south to Somalia to seek the source of frankincense (*Boswellia sacra*) and myrrh (*Commiphora myrrha*). Trees the plant hunters discovered were dug up, brought back to Thebes, and planted.[1] When he laid siege to Rome 2,500 years ago, the Gothic leader Alaric demanded a ransom, not just of gold and silver, but 1,364 kilograms of pepper (figure 7.1). Kings and queens underwrote the voyages of Marco Polo, Ferdinand Magellan, and Christopher Columbus to find faster routes to spice-growing countries. Spices have played a major role in human history.

We are the only species that manipulates its food for taste. We find foods more appealing if they contain pungent plant products, but our attempts to manipulate and control the food we eat go well beyond flavor. Food is a rich source of metaphor. We may describe a person as being "sweet" or a "sourpuss." We speak of "getting to the meat of the matter" or "where's the beef?" Many cultures have elaborate rules governing how food must be prepared if it is to be acceptable for human consumption. Orthodox Jews and Muslims have extensive dietary restrictions, but dietary restrictions also govern the diets of observant Buddhists, Seventh-day Adventists, Hindus, Roman Catholics, and Mormons. We must eat to live, but why is eating surrounded by such a complex set of rules and responses? Why do we socialize over food, celebrate eating, and spend so much time eating? Why does food become both a mark of status and a general expression of a civilization?[2]

Perspectives from evolutionary biology can help us understand many of these seemingly peculiar aspects of our relationships with food. Food

Figure 7.1. Harvesting fruits of a pepper tree, Java, Indonesia, from
La Cosmographie Universelle, 1575, by André Thevet.

selection, after all, is one of the two most important forces in animal
evolution. When an ecologist encounters a new species, the only question more important than "what does it eat?" is "what eats it?" In chapter 5, we discussed our responses to who eats or attacks us. Here we focus
on what we eat, why we eat it, and the complex emotions that surround
seeking and eating food. To understand what we eat and why we like it,
we need to probe into our coevolution with the species we eat, use as
medicines, and appreciate aesthetically.

We've seen that assessing an environment's current and long-term potential to provide food is an important task during the explore phase of habitat selection. Sometimes that potential is obvious, as when a glimpse of an unfamiliar landscape reveals herds of large mammals, flocks of birds, and fruit-laden trees. More frequently, however, evidence for food availability is less obvious. Evidence of abundant food today may not predict abundance in the future when we will need it. Once humans settle in a particular environment, we consistently manipulate it to improve its food-producing abilities (establish). These habits began long before the invention of agriculture, which is the most important way we alter landscapes today. Our preagricultural ancestors burned vast areas to stimulate new grass growth and affect distribution of large mammals. They dammed and diverted small rivers to facilitate catching fish.

Our Ancestors' Diet

Today when humans adopt a vegetable-based diet we do so by choice, but this was not always the case. Our common ancestor with chimpanzees, our closest living relatives, was probably an herbivore, as gorillas are today. Although it is unclear whether our ancestors ate other animals before they began to walk upright, by 1.6 million years ago improved bipedal locomotion made them better at capturing mobile animals. At first, our ancestors probably captured animals by grabbing them with their hands, a method still practiced in some parts of the world. The Maoris of New Zealand, for instance, construct dams to create artificial dead-end creeks into which they chase fish. When the stream flow is re-diverted they capture the stranded fish by hand (figure 7.2). On California beaches where pole fishing is illegal, people capture spawning grunion by hand. During a Nigerian fishing festival in the Sokoto River near Argungu, the first prize is awarded to the angler who captures the first fish with his hands.[3]

Killing at a distance probably began with throwing rocks at small birds, mammals, and fish. Humans are remarkably good at throwing objects with great accuracy. Chimps also throw objects using the same eighty-eight muscles that we use, but they have terrible aim; their brains evidently lack the capacity to execute the motion. The so-called Acheulian hand ax may not have been an ax but a throwing stone. Neurophysiologist William Calvin pointed out that the stones have sharp edges where they would have been held. If they had used them as tools,

Figure 7.2. Two Maori men at work with two *hīnaki* (fish traps) capturing *upokororo*, a New Zealand fish resembling a grayling or trout. This photo was taken in 1922 by James McDonald.

people would have cut themselves. They are found in very large numbers around water holes, places where game would have concentrated. In addition, when thrown the "ax" turns over in flight, landing point down in the soil or a body of an animal. In short, the "ax" was well designed as a throwing stone (figure 7.3).[4]

Killing distance further increased with the invention of spears, bows and arrows, boomerangs, and guns. Gradually humans became the most effective predator in Earth's history, expanding our diet and causing the extinction of large mammals on several continents. A by-product of domesticating and storing the seeds of grasses and incorporating large, difficult-to-capture animals is that we acquired an energy-rich diet that could be consumed in a short period of time, at meals.

Between 2.3 and 1.8 million years ago, hominids were restricted primarily to areas in Africa with water, shade, and rocky outcrops. Their weapons probably had an effective range no greater than thirty feet. Hunting success would have depended on stalking skills and vast knowledge about habitats and behavior of animals, how fast and far they could run or fly, and their escape strategies. And their mostly hairless bodies enabled them to pursue and capture large mammals during the heat of the day as San hunters do today.

Similarly early hominids would have benefited from an ability to assess whether a dangerous animal was more likely to attack rather than flee.[5] The complex skills required to hunt successfully explain why men in modern hunter-gatherer societies do not reach their maximum hunting ability until they are thirty, even though their strength peaks at age eighteen.[6]

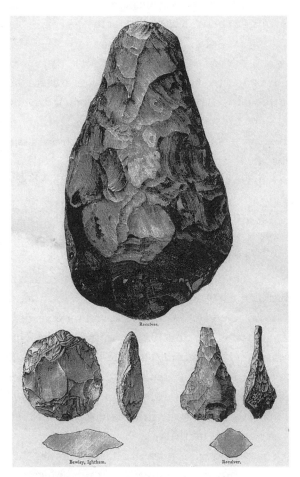

Figure 7.3. Acheulean hand axes from Kent, from the *Victoria County History of Kent*, London, 1912. The axes are named after the type-site in Saint-Acheul, near Amiens, France.

Seeking and Finding Food

Foraging—looking for food—has evolved to be emotionally satisfying for the simple but powerful reason that those of our ancestors who enjoyed foraging would have been more motivated to hunt and to take risks when attacking dangerous prey than men with lower motivation. Strong positive motivation should lead to better hunting success. Also, a successful hunter, by virtue of his success and willingness to share food with others, receives social and sexual benefits, thereby reinforcing the intrinsic pleasures of hunting.[7] Hunting and fishing are popular today even though comparable food is available at much lower prices in stores. Many people who do not like to eat fish nonetheless enjoy fishing! Catch-and-release fishing is a rapidly growing sport.

Decisions about when to forage, what to seek, and where to seek it are intricately connected because what foragers are looking for influences where they go: wild plums and crayfish along the riverbank; clams and seaweed along the strand. The foragers' goals also determine which search images they employ. We are all familiar with search images from our own behavior. For example, when I walk in the woods with an entomologist I see and hear many birds that my companion fails to observe, but I miss most of the interesting insects hiding on leaves and in bark crevices. Food reveals itself to the prepared mind.

Animals Provide Direct and Indirect Clues

Because animals move around, hunters must locate them before they can begin to stalk them. Animals leave clues in the form of tracks, feces, broken and browsed vegetation and bark, and remains of kills. These clues may tell a hunter the direction an animal was moving and how long ago it was there. The tracking abilities of hunters in traditional societies, such as the San, !Kung, and Hadza, are well documented. They can distinguish tracks of many different species and tell how long ago the animals made them.[8] I have been amazed at the tracking abilities of guides during my safaris in Africa

Matthew Sharps and colleagues wondered whether the skills and predispositions for processing animal tracks might reside in the basic architecture of our brains. If so, even modern people who do not hunt should recognize and remember animal tracks more readily than they remember other types of visual stimuli. To test this hypothesis, they asked subjects to learn and recall one hundred items, twenty from each of five categories: military armored vehicles, seashells, kitchen utensils, trees, and animal tracks.[9] Participants were informed that they would see a number of pictures and that they were to remember them because they would be asked about them later. It was a so-called double-blind study; neither the subjects nor the person administering the experiment were aware of its purpose. During the first trial, they were shown images for five seconds each, separated by a two-second interval. After a ten-minute period, during which respondents solved an arithmetic problem, they were again shown the same items. During this second trial, they saw each item again for eight seconds and were asked to write down its name. Participants remembered kitchen utensils best, but they remembered animal tracks better than seashells, trees, or military vehicles.

Sharps removed some items that confused people and randomly removed others to get seventeen items in each category. Then he conducted a second experiment with a new group of subjects. Participants best recalled cooking utensils, but again they recalled animal tracks better than seashells, trees, and military vehicles. A third experiment, using the same images but presenting them in a different order, yielded similar results. Men and women performed the same.

These findings do not suggest the existence of some sort of track-specific visual neural module. Nevertheless, Sharps's experiments suggest that we may be predisposed to pay attention to animal tracks and to recall them better than other types of unfamiliar objects. Women and men may score alike on the test because evidence suggests that Paleolithic women trapped and hunted small game. Further research will be required to clarify the mechanisms by which those responses operate, but these experiments demonstrate that such mechanisms may exist and that they can be investigated experimentally.

Plants Provide Clues to Food Availability

Few of the many plants we eat have edible tissues throughout the year. We can save valuable time by remembering when different plants flower and have ripe fruits and tender leaves. Let's see if we do and what clues we use.

We eat leaves of some plants, so being able to tell edible leaves from inedible ones is useful. Young leaves of many species, which are generally more palatable than old leaves, differ in color from mature leaves. Expanding leaves of most temperate-zone plants are a yellowish green; they turn darker green when full-sized. In contrast, expanding leaves of many tropical woody plants are reddish. Leaves of temperate zone deciduous trees assume bright colors when chlorophyll is withdrawn prior to their being discarded in autumn. Paying attention to leaf color may have made it easier for our ancestors to locate plants of particular species that may now, or in the future, have nutritious tissues.

If our ancestors paid particular attention to leaf colors and used them to find edible plant tissues, we would predict that plant breeders, wishing to sell more plants, should select for markets mutants with leaves with contrasting colors at times of years when leaves of most plants are green. Cultivated trees, such as maples, that produce neither striking flowers nor edible fruits are good for testing the prediction that atypical leaf colors have been commercial successes.

We can test this prediction by looking at the 133 cultivars of twenty species of maples illustrated in Antoine Le Hardÿ de Beaulieu's *An Illustrated Guide to Maples*.[10] Fourteen species have cultivars with unusual leaf colors. Ten of those species have brightly colored (mostly red) petioles and shoots. Mature leaves of wild individuals of the widely cultivated Japanese maple *Acer palmatum* remain green throughout the summer. But only fifty-seven of 140 listed cultivars of this species have green leaves. Sixty-six have either red or purple summer leaves; seventeen have variegated leaves.

Horticulturalists clearly have preferred mutants with leaves that are not green when they "should be." The hypothesis may also explain why we are attracted to the autumn colors of deciduous forests of northeastern North America, even though these colors signal a time of decreasing, not increasing, resources. English gardeners were particularly attracted to American plants with red and orange branches and colorful blossoms and berries.[11]

Flowers Provide Clues

Flowers are a minor part of human diets, but flowers are an important source of information about current and future locations of foods.

Watching and following bees visiting flowers may help us find honey. Flowers precede fruits and seeds; we can predict where and when we can collect those foods if we remember when and where plants bearing them bloomed. Flowering plants may also signal the presence of water. In species-rich vegetation, flowers greatly aid in identifying plants that otherwise may be difficult to distinguish in the mass of similar green leaves.

If it were advantageous for our ancestors to pay attention to flowers, we should enjoy looking at them. We do! Women smile when presented with a bouquet of flowers; their pleasurable feelings last for several days.[12] Men and women in elevators respond more positively to flowers than to other gifts. Giving flowers to elderly people evokes immediate and long-term positive moods and improved memory. Florists report that they receive hugs and kisses when they deliver flowers.

Once we have located food, we need to decide whether or not to eat it. Maybe it is not worth ingesting. A normally palatable item might be dangerous to eat if it is infested with pathogens or parasites. Thus whether or not to eat something becomes a crucially important question. As Gary Paul Nabhan notes, "We are what our ancestors ate, and also what they had to regurgitate."[13]

Danger at the Dinner Table

The act of eating breaches the body's immune defenses. This is true whether or not we're allergic to or intolerant of lactose or gluten, shellfish or peanuts. Every animal that eats takes in potential toxins, pathogens, and parasites. The liver—an evolutionary innovation also present in invertebrates like mollusks and crustaceans—is able to neutralize many toxins, but vomiting and diarrhea are humans' most effective and immediate defenses against a swallowed pathogen. Neither is pleasant, and they may be unhealthy. Thus an animal benefits by being able to distinguish edible from inedible items and ingesting only edible ones.

Getting enough food without becoming sick is especially difficult for omnivores.[14] As the name suggests, omnivores typically eat many food types; what is available varies geographically.[15] Tissues of most plant species contain a range of alkaloids, terpenes, and phenolics that are toxic to herbivores and pathogens. The search to determine which toxins plants contain and how to neutralize them has preoccupied human societies for many millennia.[16]

Ingesting an unfamiliar food is risky, yet an omnivore needs to sample many unfamiliar foods to find out which are safe to eat. Otherwise we will reject foods we should have eaten. A varied diet is likely to be more nutritionally balanced and to have lower levels of any particular toxin than a diet composed of only a few species. Thus omnivores are curious about new foods, but approach them with healthy suspicion.

When a novel food source is encountered, an animal must first determine whether or not it is edible. Then the animal must decide whether an item known to be edible should be captured and eaten at that particular moment. Finally, if the animal is provisioning offspring, it needs to decide whether or not to eat the item immediately, carry it back to its young, or cache it. The best decision varies during the course of a year or an animal's reproductive life. Depending on age and whether the animal has a mate or young, it may store food for future consumption or deliver it to be shared by a mate or kin.[17] Different considerations and emotions accompany these decisions.

Is It Edible?

We eat hundreds of different species, but we reject millions of others. Potential foods are not simply either edible or inedible. Some are highly nutritious and always desirable. Others provide specific nutrients that are only occasionally needed. Some are low quality, starvation foods. During times of scarcity, foods that are otherwise rejected may be eaten. As humans deplete their food base, they turn to increasingly less profitable and palatable foods.[18] Charlie Chaplin making a seven-course meal of a shoe hints at the real starvation foods humans have turned to in lean times: during the Irish potato famine, starving Irish ate kelp and seaweed. During World War II, the Dutch in occupied Holland ate tulip bulbs; in Russia, peasants learned to survive on nettle soup. Even cannibalism may be resorted to under extreme duress.

Omnivores are rarely born knowing what to eat. As Paul Rozin put it, "it's part of the biology of being an omnivore to have to learn almost everything about what is edible."[19] We learn what is edible by carefully ingesting things and monitoring the physiological consequences, by observing what others eat, and by being told what is good to eat. Observation and instruction have obvious value; individuals we observe have probably ingested those substances for many years and have survived. Mother may indeed know best!

Breast milk has both immunological and antibiotic properties, so nursing infants can sample unfamiliar objects with less risk than after they are weaned. A relatively safe way for children to learn what is edible is to sample things within view of caretakers. Very young children are more likely to be in the presence of caretakers, but when children are able to walk they are more likely to sample foods where adults cannot see and advise them.

It follows that infants should begin to put strange objects in their mouths as soon as they are able to but reduce doing so when they wean. As expected, infants begin to put objects into their mouths at about two months of age, before they begin to crawl. At this age, we use objects inserted into the mouth to pacify crying infants or to provide visual interest.[20] Mouthing increases up until about six months of age and then begins to decrease. It is rarely observed in children older than two years.[21] By mouthing small objects young children also ingest helpful microbes that will replace those in breast milk when they are weaned.[22] Flavor preferences develop rapidly in newborn infants. Several feeding experiences are generally required to develop flavor preferences, but aversions are learned more rapidly.[23]

Without instructors, how can an omnivore tell whether ingesting a food item would be dangerous? For most types of objects, we immediately can tell if it is dangerously hot or cold, sharp or irritating, soft or hard, bitter or sweet. But if we ingest something toxic, we do not get sick until hours later. Like other omnivores, we automatically and unconsciously associate sickness with food we ingested hours earlier. We develop aversions to that food, and typically avoid it in the future. Even odor of a food that made us sick may evoke nausea. I know this from experience. As a youth I became ill from eating too many clams on my first visit to the Atlantic coast. I was unable to eat clams again for about twenty years. Many of you have doubtless had similar experiences.

Many different responses to potential foods, ranging from extreme pleasure to disgust, have evolved in response to the danger of ingesting external objects, even nutritious ones. People who ingest contaminated or spoiled food are likely to become ill or die. It is a good idea to pay attention to signs that a potential food might be toxic, even when one is very hungry.

A Matter of Taste

We may accept a food because it tastes good or reject it because it tastes bad, but that is only one of many reasons we do so. We also decide on the basis of anticipated consequences, both physiological and social, of eating a type of food. What you know about a food—where it comes from, its characteristics—may influence whether you think it is appropriate to eat. Foods may be acceptable only in particular situations, or they may be categorically rejected. Because of the variety of reasons that influence which foods are actually acceptable, human cuisines differ more than expected from differences in what is available.[24] Nonetheless, a few generalizations apply.

Pleasure is positively associated with nutritional value. As we saw in chapter 4, we like the taste of nutritious foods. Under ancestral conditions, sugar-rich foods, such as fruit and honey, were scarce. We cannot ask a biochemist to tell us where in the structure of a sugar molecule sweetness lies. Sweetness comes from an interaction between the structure of a molecule and our nervous system. Our nervous system has evolved to render simple sugars, which are easily digested and energy rich, as sweet. We have used knowledge of how our neural receptors function to develop nonnutritious sugar substitutes that trigger sweetness sensations so that we can enjoy artificially sweetened food and drink while avoiding unwanted calories. Until our ancestors added animals to their diet, they also found it difficult to acquire fat.

Disgust

Charles Darwin was the first person to identify the strong relationship between disgust and food. In *The Expression of the Emotions in Man and Animals* (1872), he remarked, "Disgust is a sensation rather more distinct in its nature, and refers to something revolting, primarily in relation to the sense of taste, as actually perceived or vividly imagined; and secondarily to anything which causes a similar feeling, through the sense of smell, touch, or even eyesight. Nevertheless, extreme contempt, or, as it is often called, loathing contempt, hardly differs from disgust"(250). A few pages later, Darwin further noted, "It is curious how readily this feeling is excited by anything unusual in the appearance, odour, or nature of our food" (255). He also commented, "Extreme disgust is expressed by movements round the mouth identical with those preparatory to the act of vomiting. The mouth is opened widely, with the upper lip strongly retracted, which wrinkles the sides of the nose, and with the lower lip protruded and everted as much as possible" (256). "Spitting seems an almost universal sign of contempt or disgust; and spitting obviously represents the rejection of anything offensive from the mouth" (259).[25]

Darwin's pioneering observations have been confirmed. Disgust, which is expressed in all human cultures, is the most powerful reason people reject certain types of food. We feel disgust toward objects our culture tells us are not food. Most items that evoke disgust are of animal origin. Most of us are sure that disgusting things, such as worms, rats, and beetles, taste bad even though we have never tried them. Moreover, we have no interest in trying them. Pus, maggots, rotting food, and scavenging animals universally evoke the distinctive facial expressions of disgust (figure 7.4). We respond with disgust both to things that should be kept out of our bodies and to things, such as blood, that should be kept in. Disgust carries with it a notion of contamination. Most people will not drink juice that has been stirred with a sterilized cockroach or drink from a meticulously cleaned bedpan. In most cultures, people eat relatively few of the available animals. Americans generally do not eat insects, amphibians, reptiles, or rodents, but people in some other cultures do. Yet, most kinds of animals that evoke disgust are actually edible and nutritious!

Darwin did not speculate about the origin of feelings of disgust, but his observations fit with the notion that disgust is an adaptation that deterred our ancestors from eating rotting meat, feces, and viscera that commonly harbor harmful microorganisms. Microorganisms can multiply rapidly; there is no dose below which it is safe to ingest them. Our responses to potential contamination are, appropriately, dose insensitive. Even brief contact with a potential contaminant suffices to evoke full-fledged feelings of disgust.[26] Disgust is evolutionarily programmed

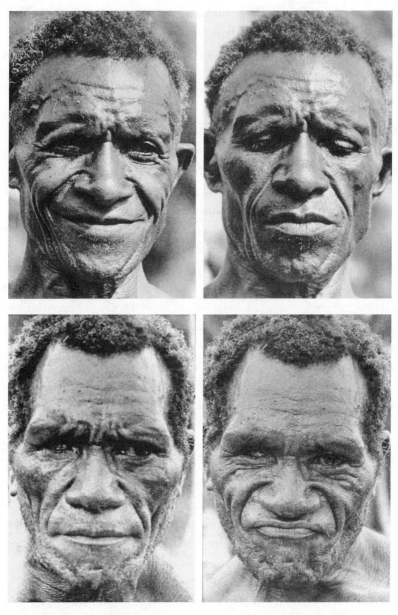

Figure 7.4. In 1968, psychologist Paul Ekman took his camera to the island of New Guinea to photograph the faces of the South Fore people. These four photos show a man smiling, looking down, concerned, and frowning.

intuitive microbiology that developed long before people knew that microorganisms existed, much less that they caused diseases.

Inappropriate items are ones our culture tells us are inedible. We do not eat things like pencils, grass, paper, and cloth even though we probably would not get sick if we did. Plant parts, which are generally less nutritious and more difficult to digest than animal tissues, rarely evoke disgust. They are also much less likely than animal tissues to harbor toxic bacteria. Many vegetarians, particularly those who avoid eating meat for moral reasons (saving animal lives, reducing animal pain, saving the environment), come to dislike meat and may even be disgusted by the thought of eating it. Vegetarians who avoid meat primarily for health reasons generally remain tempted by meat and avoid exposure to their favorite meat dishes.[27]

Our sense of purity may also have intuitive microbiological roots. Our minds appear to have a "purity module" that directs our attention to things that were associated with dangers caused by microbes. Concerns about purity have simulated development of elaborate, complex moral rules that in many cultures govern diet and hygienic practices.[28] Humans have universally strong moral feelings about menstruation, eating, bathing, sex, and handling corpses. Much of the moral law of Judaism, Christianity, Hinduism, Islam, and many traditional societies functions to maintain "purity." People who observed those morally enforced practices may have experienced better health than those who did not. Compliance was probably maintained because violations of those rules evoked disgust.

All cultures and languages studied by Jonathan Haidt have at least one word that applies to both core disgust (cockroaches and feces) and to some kinds of social offence, such as sleazy politicians or hypocrites.[29] We reject or disapprove of abstract concepts such as ideologies and political views with disgust. Propagandists throughout history have exploited this connection to evoke disgust toward other groups of people. Nazi propaganda depicted Jews as cockroaches and rats; Hutu instigators labeled Tutsis as cockroaches during the Rwandan genocide.

The *levator labii* muscle region of the face, a part of the basic oral/nasal disgust response to foods, is activated by moral disgust as well.[30] This suggests that our sense of moral disgust may be built upon an ancient food rejection system.[31] When we say that moral transgressions "leave a bad taste in my mouth," we may be unconsciously expressing a deep evolutionary relationship.

Our intuitive microbiology may also account for the remarkable fact that animal tissues are both the most tabooed and the most preferred types of food in most human cultures. These responses make sense because animal tissues are highly nutritious but are fertile breeding grounds for pathogens once the animal dies. We have strong biological urges to eat animals, but we also have a rich array of cultural reservations and rules about eating them. Part of our reluctance to eat animals may come from a belief that "you are what you eat," that you take on some properties of things you ingest. Many people hold, either consciously or subconsciously, a belief that if you eat animals, you will become more animallike. Such beliefs fueled opposition to vaccination. Injecting fluids from cows into our bodies, it was feared, would animalize us.[32] Yet, people are not concerned about becoming more plantlike as a result of eating plants.

Accepting New Foods

We may eat hundreds of species of plants and animals, but human diets are remarkably conservative. We are reluctant to try new foods or to abandon familiar ones. Europeans were reluctant to add potatoes and tomatoes to their cuisines even though they knew that those plants were staple foods of Amerindians. Given such resistance, how do new foods become incorporated into cuisines? Spices play a critical role in determining what is acceptable; once established, they are remarkably persistent.

Some Like It Hot

Spices influence tastes of foods, but they contain little energy. They must provide some other benefit for us to value them so highly. One hypothesis is that we enjoy spices because they disguise the tastes and smells of spoiled food. However, food-borne bacteria kill thousands of people every year and debilitate millions more. Eating spoiled food by covering up its bad flavor is a dangerous thing to do, even if you are starving. Natural selection is not likely to have favored people who ate rancid food whose bad flavors were covered by spices.

An antimicrobial hypothesis is more plausible. Spices protect plants against bacteria and fungi. They also protect our food from attack by them. Many of the most widely used spices have strong antimicrobial properties.[33] Mixes of them, which are common in many traditional recipes, are even more powerful. Disease-causing organisms are more abundant in tropical than in temperate regions. As we would expect, the proportion of traditional recipes containing antimicrobial spices is highest in the tropics.

Cells of dead plants harbor fewer pathogens than cells of dead animals. An animal's immune system ceases to function when it dies; bacteria then increase rapidly. The antimicrobial hypothesis therefore also predicts that spices should be used less in vegetable dishes than in meat dishes. The prediction has been confirmed. As shown by more than two thousand recipes from more than one hundred traditional cookbooks from several dozen countries, vegetable dishes everywhere call for fewer spices per recipe than meat dishes.[34] Once people learned that they were less likely to get sick after eating spicy foods than nonspicy foods, spices became signals that a food was safe to eat.

Pathogens and foods available to people vary geographically, so people rapidly evolved genetic adaptations to deal with the peculiarities of their local foods.[35] People who now live in areas remote from those occupied by their ancestors may actually benefit from devoting financial resources to import the spices that characterize their traditional cuisines.

Why are we the only animals that add spices to their food? When did we first begin to use spices? Nobody knows the answer to either question, but spices would have become especially valuable when our ancestors began to hunt and kill large mammals whose flesh could not be eaten quickly. Humans may have been using spices and salt to preserve food for millennia.[36] The value of spices might have further increased when people learned how to cook food. Cooking neutralizes many toxins, allowing us to eat plants that would otherwise be inedible. Apes use plants and minerals as antacids and purgatives and worming medicine. When they are sick, chimpanzees seek out and eat bitter-tasting plants that they otherwise do not ingest.[37]

Many factors have been proposed as key to the rapid recent evolutionary changes in the human lineage. Among them are bipedal locomotion, opposable thumbs, language, trade, and a more complex social organization. Control of fire has also been suggested as a critical factor—the so-called cooking hypothesis. Control of fire provided protection from predators and the ability to shape metal and fire ceramics as well as cook food. Cooking shifted much of the work of chewing and digesting outside our bodies. Nonhuman primates also prefer cooked over uncooked food.[38] According to the theory, by "outsourcing" part of the digestive process we freed up energy that became available to support a larger and energetically expensive brain.[39]

What Foods Are Eaten Only at Particular Times?

Many nutritious foods are eaten only during special occasions. Foods that are suitable at certain times may not be suitable at other times or under other conditions. Some of these restrictions are purely cultural, but one has a clear biological basis.

About 70 percent of women experience nausea and vomiting (pregnancy sickness or "morning sickness") during the first trimester of pregnancy. Morning sickness has traditionally been viewed as pathological; efforts were made to suppress it. However, Margie Profit suggested that morning sickness is an adaptation that discourages mothers from ingesting toxins that may be harmful to a developing embryo.[40] Strong aversions to certain foods are confined to the first trimester of pregnancy, the time of major organ formation, when a developing embryo is most vulnerable to toxins. Foods that most frequently evoke aversions or sickness contain high levels of toxins (coffee, alcoholic beverages, strong-tasting vegetables, and sometimes meats and eggs). Women who suffer pregnancy sickness have lower probability of miscarriage and fetal death rates than women who do not. Morning sickness is absent in all seven of twenty-seven traditional societies with meat-free cereal-based diets.[41] If morning sickness is an evolved adaptation, it probably evolved recently, after our ancestors left Africa, developed agriculture, and added many more foods to their diets.

Food and Social Status

Food sharing has ancient roots; the practice is widespread among hunter-gatherers. Males of many species of birds and mammals court females using food. Male chimpanzees show off by sharing valuable and difficult-to-capture food—for example, part of a red colobus monkey.[42] Food sharing had the added benefit of making sure meat was eaten before it could rot, necessary in tropical climates where meat quickly rots and preserving it by smoking and drying is difficult. And once we began sharing food with one another, we began to use food to show off. A difficult-to-capture food item is good for showing off with because it is genuinely valuable and clearly signals the skills of its possessor. (figure 7.5).

In traditional hunter-gatherer societies, successful hunters advance in social status and have enhanced reproductive opportunities.[43] Complex networks of obligations govern food sharing. A prize kill like a kudu antelope means a bonanza of meat—everyone gets a generous portion. The items most likely to be shared are hunted rather than gathered.[44]

Figure 7.5. Italian truffle hunter and trader Cristiano Savini, *left*, holds a 1.5-kilogram (3.3-pound) white truffle; Angela Leong, *right*, wife of Macau tycoon Stanley Ho, won the auction for the truffle with a bid of US$330,000 in Macau, December 1, 2007. The unusually heavy truffle was dug up in 2007 by Savini, his father, Luciano, and dog, Rocco, in Palaia, a town about forty kilometers (twenty-five miles) from Pisa. The Savinis said Rocco started sniffing "like crazy" when he zeroed in on the fungus.

Foods that are of high value and difficult to obtain, especially meat, are shared more widely than easily acquired foods. Aché hunter-gatherers share meat and honey within their band, but they share vegetable foods, which depend mostly on effort expended rather than skill, only within nuclear families.[45]

Good hunters in hunter-gatherer societies on average share more food with others than do poor hunters. A good hunter might do better for his family by hunting smaller, more reliable, and less shareable resources, but good hunters and their families receive preferential treatment within the group. Their legitimate offspring survive better than offspring of poor hunters. Good hunters also father more illegitimate children; women report that they prefer them as lovers.[46] Not surprisingly, hunters preferentially seek resources they can share and use to gain reproductive benefits. Show-offs tend to spend more time away from their spouses, thereby increasing opportunities for gaining sexual favors.[47]

Men typically show off with difficult-to-capture food, but today they also compete with superlarge vegetables, even though unusually large ones are less edible than smaller ones. People may try to convince others that their gardening success comes from having personal magical abilities.[48]

Domestication: Making Foods Better

As we have seen, features of prey that signal their suitability as food evoke positive emotional responses. Our ancestors domesticated plants and animals by enhancing traits that made them even more profitable to eat. Animals were bred for larger size, more meat, and docility. Domesticators favored tenderness and lower concentrations of bitter-tasting toxins in plant leaves. In fruits they increased size, sweetness, pulp-to-seed ratios, ease of separating seeds from pulp, and storability. Domesticated flowers are larger and more conspicuousness than their wild relatives. Some plants that were toxic in their wild form, like cassava, were crossbred to create nontoxic varieties.

Fleshy fruits evolved to attract animals that consume pulp and then defecate unharmed seeds away from the parent plant. Most fleshy fruits are nutritious, but many contain toxins until their seeds are mature.

A few fruits are toxic even when mature. Among them are chili peppers, which contain chemicals (capsaicins) that irritate the linings of our mouths. These irritants have been bred out of sweet peppers, but hot peppers have been incorporated into cuisines throughout the world, probably because of their antimicrobial properties.

Energy allocated to pulp is not available to be used to produce seeds. Plants evolve to allocate only enough resources to pulp to attract frugivores. Frugivores may prefer larger and more nutritionally concentrated fruits, but they exerted little influence on the evolution of fruit size and richness. Fruits were too scarce during many times of year for frugivores to be choosy! Domestication dramatically changed all that. Fruits of domesticated varieties of apples, pears, peaches, plums, quinces, apricots, oranges, grapefruits, and others are much larger and have greater concentrations of sugar than those of their wild ancestors. We have even produced many varieties of seedless fruit-bearing plants that we must propagate by grafting. We gain because the plant allocates none of its energy to seeds we don't want.

Calling for Food

People generally stalk animals quietly, but sometimes people sing when they hunt; they may imitate the calls of animals to lure them into killing range. As we saw in chapter 1, hunters may sing for honey. We are remarkably vocal animals. In the next chapter we will explore how we came to be so talkative and musical.

8

The Musical Ape

The first song of a male red-winged blackbird in late winter evokes in me feelings of great pleasure. That distinctive *conk-a-ree* reminds me of the many happy hours I have spent among these birds, studying their social lives and trying to tease apart the meaning of their alarm and contact calls and songs (figure 8.1). I have watched them call and sing to attract mates, repel competitors, and warn others about imminent danger as a hawk flies by. I tried to find out what messages the singing males were communicating and how other individuals responded to the messages.

I learned that songs tell other males that the territory is occupied and that its owner will attack them if they trespass. Females, on the other hand, are informed that the singer will welcome them, will fertilize their eggs and help them defend and rear their offspring. Female redwings also sing, but they do not attract mates or defend a territory. Redwings were clearly responding emotionally to their calls and songs, but I knew that what they felt was hidden from us.

In part because I carried out most of my research in the field, I am sensitive to nature's sounds. In this way I resemble members of hunter-gatherer and traditional agricultural societies, but they live close to nature day and night, so they know more about nature's sounds than I do. For example, the Hutu and Tutsi tribes of central East Africa knew the ultralow frequency communications of elephants and incorporated them into their songs and stories for centuries before modern scientists became aware of them. For millennia, Tlingit hunters in the Pacific Northwest and Inuit hunters in the Arctic heard the sounds of whales through the hulls of their boats. Scientists first recorded them in the 1940s.

Figure 8.1. A male red-winged blackbird (*Agelaius phoeniceus*) utters a
song while he spreads his wings and exposes his red shoulder patches.
This song-spread repels other males but attracts females.

Sounds have been a source of valuable information throughout human history. Young BaAka children in the Dzanga-Sangha forest of the
Central African Republic know what many of the sounds in the forest
mean for food and danger.[1] This chapter explores the origin and subsequent development of our remarkably powerful emotional responses
to sounds. I ask, and hope to answer, two questions. First, how did paying attention to natural sounds and sounds produced by fellow humans
improve our ancestors' survival? Second, how did the benefits of paying
attention to the sounds of nature lead to the elaboration of music both
by our voices and by musical instruments? We differ strikingly from our
closest primate relatives in being remarkably musical.

To understand how our emotional reaction to sound helped us survive and may have led to the development of music, we need to understand the acoustic environment of the natural world. A soundscape consists of a complex web of inorganic sounds produced by the physical
environment and organic sounds produced by living things. Much like
the instruments of an orchestra, the sound of each component has its
own frequency, amplitude, timbre, and duration. Taken together, these

components make up the acoustic ambiance of a given habitat. Important messages may be conveyed by the "orchestral" ambient sound of a forest or meadow.[2] A meadow where many individuals are calling may signal an environment free of predators. Conversely, the sudden cessation of calls probably indicates a significant change. In the natural world, sudden silence is a real attention grabber.

Many of nature's sounds come from the physical environment. Waves crash, thunder rolls, volcanoes erupt, avalanches roar, and brooks babble. Many physical events of significance to animals, such as electrical storms, high winds, volcanic eruptions, and falling water, produce loud noises. Other sounds, such as the movement of air through tree canopies, grasses, or dry leaves, are softer.

In the beginning, living things produced sounds only as a by-product of metabolic processes, the business of living, or by moving. The neural circuitry at the base of the midbrain and upper spinal cord that underlies sound production was present more than four hundred million years ago in the common ancestor of all vocal animals, but animals probably did not produce other than incidental sounds until more than three-quarters of life's evolutionary time had passed. Some insects, fish, and reptiles and most amphibians, mammals, and birds call and sing, but even today, most animals are mute. Many silent animals can detect and respond to sounds and other vibrations in their environment. Flying moths initiate evasive actions when they hear the calls of hunting bats.

Using the channel of sound for communication was an important innovation. It allowed animals to send messages without revealing themselves to potential predators. A visual signaler makes itself conspicuous, but the source of a sound can be difficult to pinpoint because sounds bend around corners and penetrate obstacles. Sound is also a fast and efficient way of sending signals. An animal can pass in an instant from uttering a high to a low pitch, from harsh notes to pure tones, and from loud to soft notes, from calling to silence. An animal can stop displaying, but it does not instantly disappear. Many animals, especially small ones like insects, amphibians, and birds, can broadcast a message from a sheltered location or at night when they would be difficult for predators to see.

Attending and responding to the sounds of nature has survival value for both vocal and nonvocal animals. Paying attention to nature's sounds certainly had survival value for our remote ancestors. The sound of rushing water may indicate the presence of dangerous rapids ahead. Sounds may reveal the presence of desirable prey, dangerous predators, or hostile human enemies. The buzzing of bees may indicate the location of honey. Hunters who attended to animal sounds would have been more successful than individuals who ignored them.

Some birds and whales elaborate their calls into songs that we call musical. They imitate one another and compete for space and mates by means of singing duels. Many birds learn their songs by listening to others; they may form dialects. Yet, the calls of most primates are relatively simple and innate. Humans differ strikingly from our closest relatives in our elaborate music making. What led to that remarkable development?

All Human Cultures Make Music

Everyone can sing and has songs inside. If you don't think you can sing, go into the forest and start singing. Pretty soon the trees will start to respond—swaying and moving their leaves.

—A Menominee Elder

Making music is ancient and ubiquitous. No human culture known today or at any time in recorded history lacked music. Music generates powerful emotions of pleasure, ecstasy, joy, sadness, terror, and disgust. Music exerts a powerful influence on our minds and bodies. Music therapy has been widely used in many societies since antiquity.[3] Many hospitals and clinics employ music to calm patients and reduce the need for general anesthesia. Premature infants gain weight faster and leave the hospital sooner if they can listen to lullabies.[4]

It is remarkable that something as unnecessary as music should exert such an influence over our lives. The powerful emotions music evokes in us suggest that individuals with those emotions had better survival and reproductive success than individuals with weaker emotional responses. But why?

The origin and function of music have attracted the attention of scholars for more than one hundred years. During the first half of the twentieth century, the field of ethnomusicology, the study of social and cultural aspects of music and dance in local to global contexts, flourished. Its practitioners wanted to understand the origins and evolution of music and the place of music in the human mind. They applied theories from many social science fields to interpret the forms and meanings of music worldwide. The International Council for Traditional Music was founded in 1947; the Society for Ethnomusicology was founded in 1958.

An interesting example of ethnomusicology research is the Cantometrics Project. Launched in 1959, its goal was to test Alan Lomax's theory that the musical style of a culture, that is, the pattern of its common musical practices, reflects its social organization and lifestyles. With his collaborates, Lomax analyzed more than four thousand songs from more than four hundred cultures. His major conclusion, published in a book,[5] was that "a culture's favored song style reflects and reinforces the kind of behavior essential to its main subsistence efforts and to its central and controlling social institutions."

The lives of people in most of the societies that Lomax studied were dominated by hunting, gathering, fishing, tending animals, gardening, and, of course, politics. Those activities were the most important things that their song styles could reflect. We cannot recover the music of our remote ancestors, but Lomax's findings probably offer valuable clues to its forms.

The Origins of Music

The first person to speculate seriously about the evolution of music was none other than Charles Darwin. Darwin's wife, Emma, was a talented pianist; her daily piano playing captivated him. Darwin's home life and his love of music clearly influenced his thinking. In *The Descent of Man, and Selection in Relation to Sex*, he devoted ten pages to birdsong and six to human music. He viewed both phenomena as outcomes of sexual selection, features that originated and became elaborated as displays to attract sexual partners. Darwin suggested that the capacity to perceive musical notes could easily have begun as a side effect of the capacity to distinguish noises "an ear capable of discriminating noises—and the high importance of this power to all animals is admitted by every one—must be sensitive to musical notes" Darwin concluded by proposing that if male birds sing to females, it must be because female birds are impressed by singing: "unless females were able to appreciate such sounds and were excited or charmed by them, the persevering efforts of the males, and the complex structures often possessed by them alone, would be useless; and this is impossible to believe" And finally, "it appears probable that the progenitors of man, either the males or females or both sexes, before acquiring the power of expressing their mutual love in articulate language, endeavored to charm each other with musical notes and rhythm."[6]

Darwin's insightful conjecture about birdsong was quickly accepted, but his suggestion that human music evolved to serve the same function was neglected until very recently.[7] Claude Levi-Strauss expressed the position typical of cultural anthropologists, as follows: "Since music is the only language with the contradictory attributes of being at once intelligible and untranslatable, the musical creator is a being comparable to the gods, and music itself the supreme mystery of the science of man." In other words, we were unweaving the rainbow again.[8]

But opinions have changed. In his 2005 book, *The Singing Neanderthals*, Stephen Mithen argues that it is highly unlikely "that our deepest emotions would be so easily and profoundly stirred by music if it were no more than a recent human invention. And neither would our bodies, as they are when we automatically begin tapping our fingers and toes when listening to music. In fact, even when we sit still, the motor areas of our brains are activated by music."[9]

In recent decades, thinkers from a range of disciplines have come forward to propose new hypotheses about how human music might have arisen. Some suggest music is a by-product of having a big brain, or of being conscious, of having developed an elaborate, learned culture, or of having surplus energy. None of these explanations made the case for music conferring any adaptive advantage. Those arguments make little evolutionary sense because surplus energy can be converted into fat or used to synthesize products the body needs and that enhance an individual's fitness. A surplus brain mass that consumed great amounts of energy but conferred no benefits would quickly have been eliminated by natural selection.[10]

Most theorists who seriously sought an adaptive value to music postulated group selection as the mechanism. In other words, they thought that music benefited social groups rather than the individuals making and responding to it. Although a group may benefit from music making by its members, theories that explain the evolution of music by postulating benefits to groups face the freeloader problem. If the production of music benefits the group but is costly to its performers, individuals that save time and energy by not incurring those costs still gain the benefits. They get something for nothing. Thus, freeloaders should increase at the expense of performers. Because the freeloader problem is serious, I limit my discussion to those hypotheses that identify benefits to individuals who pay attention to and make music.

We have seen throughout this book that behaviors that made us successful tend to be ones we find rewarding. Therefore, if paying attention to and responding to the sounds of nature contributed to the success of our ancestors, doing so should be intrinsically rewarding and the sounds of greatest potential significance, that is, those that signaled the most important changes in the environment, should generate the strongest emotional responses.

Nevertheless, there is no obvious reason that attending to and responding to the sounds of nature should lead to the development of music. Individuals of most species benefit from attending to the sounds of nature, but most of them did not develop music. Music accompanies almost everything humans do—hunting, herding, storytelling, playing, washing, eating, praying, meditating, courting, marrying, healing, and burying. Powerful selective forces must have been acting on us for us to become the most highly musical among the millions of species. To find out what they were, a brief excursion into history may be helpful.

Music's History

The physical traces of music's origin are lost in the mists of antiquity. Early vocal music has left no traces, but we have reasons to believe that singing is very ancient. In the late 1980s, French archaeologists used singing in their exploration of prehistoric caves in southwestern France. They discovered that the chambers with most paintings were particularly resonant and concluded that the caves were sites of ceremonies accompanied by music.

Instrumental music probably began with percussion—slapping the buttocks, belly, and thighs, clapping the hands, and stamping the ground. The manufacture of instruments goes back at least to fifty thousand years ago. Beaded rattles, scrapers, bullroarers, and bone flutes have been excavated from late Older Stone Age strata. The earliest known musical instruments, fragments of bone flutes from Slovenia, are about forty-four thousand years old.[11] Instruments in the Hohle Fels Cave in southwestern Germany are at least forty thousand years old; those from Geissenklösterle in southern Germany were made about thirty-six thousand years ago.[12]

Although humans have been making music for at least fifty thousand years, until the invention of musical notation in the Middle East, about four thousand years ago, there was no way to capture and preserve its forms. With the invention of musical notation we gradually came to express ourselves musically in the form of repeat performances.

We can clarify our thinking about the origins and evolution of music by focusing on several questions.[13] First, what adaptive functions are served by singing, chanting, humming, whistling, dancing, drumming, and playing instruments? Second, who generates these signals, what behavioral changes do they generate, and who benefits? Third, what are the costs of producing and listening to music? These questions require us to postulate an origin of musical utterances and to show how engaging in musical behavior enhanced the survival and reproductive success of performers and recipients, whether or not other group member also received benefits.

Costs and Benefits of Making Music

Many theorists have argued that because performing and listening to music is cost free, music, and other arts as well, might be by-products of other adaptations, such as "the hunger for status, the aesthetic pleasure of experiencing adaptive objects and environments, and the ability to design artifacts to achieve desired ends."[14] Clearly, these adaptations contribute to our enjoyment of music today, and, yes, a decision to participate in some kind of musical activity imposes little cost in modern societies. However, for three reasons making music was costly for our ancestors.

First, making or listening to music imposes opportunity costs. When performing or listening to music, one cannot perform many other activities. Singing has traditionally accompanied some kinds of work, such as rowing, plowing and harvesting, and hauling nets, in which coordination of the movements of individuals is important. We can and do sing while cooking, drawing water, and seeking food, but we can do very few other things while playing an instrument.

Second, singing, playing an instrument, and dancing consume energy. Our metabolic rate during vigorous dancing is six times greater than when we are resting. Dancing is as energetically expensive as swimming and twice as costly as rapid walking. Vigorous drumming is equally costly. Dancers in traditional societies often collapse in exhaustion. The ability to perform complex dance movements for extended time periods demands physical vigor and muscular coordination.

Third, individuals involved in making music usually do not pay careful attention to their surroundings. This would have left early human performers and listeners vulnerable to surprise attack. Music masks other sounds and advertises the location of the group. Predators and hostile humans can easily locate and approach undetected a group of music makers.

What benefits could have overridden those costs? That is, from an evolutionary point of view, what good is making music?

What Benefits Accrue from Making Music?

Making music is typically a group activity; in many cultures music is inseparable from dancing. Dance plays an important role in courtship in many cultures. Maasai warriors sing and dance to display their personal prowess before assembled maidens, using songs to extol their manly virtues.

Group dancing, rhythmic clapping, and chanting all release brain chemicals that promote affiliative emotions among both athletes and warriors. Scottish regiments marched into battle preceded by their pipers as recently as the 1944 D-day landings in Normandy. The traditional Maori haka, used to intimidate enemies in battle, has been adapted to the rugby pitch by the New Zealand national team, the All Blacks. Singing and dancing reinforce the bonds among us and arouse in us feelings of love and a willingness to act for the benefit of the group.

Song is widely used as a signal in the animal world, often in combination with striking visual displays. Many insects and amphibians communicate through song. Male birds use song to claim and hold territories and to win mates and keep them. Among the species of birds that establish long-term pair bonds, many cement those bonds by engaging in elaborate vocal duets.

Some primates use calls to defend their territory, advertise for a mate, or both.[15] They also use calls to signal the presence of predators. But these calls are at best only simple songs. We humans are the only truly singing primate. How and why did our ability to sing arise?

Theories for the Origins of Music

In addition to explaining how it would have benefited both performers and listeners, any theory of the origin of human music must explain how music became so remarkably elaborated. Some theories put forward to explain the origin of music fail to pass the test that both sender and receiver must benefit. Others falter when confronted with the high costs of making music. And most fail to explain the evolution of music's ability to evoke remarkably powerful emotions. Nonetheless, components of those theories can help us understand music's origins.

Origin from Language. Language has often been suggested as the source from which music developed.[16] Chanting, an early musical form, exaggerates the patterns, tones, and rhythms of speech. Buddhist chants are rooted in spoken language; they are governed by qualitative rules such as "keep the tones without fault."

Nevertheless, language alone cannot account for the development of music. A major objection to this theory is that we process music and language mostly in different areas of our brains. Language is mediated on the left side; music is mediated on both sides. The rhythmic component of very simple spoken songs, like nursery rhymes, is handled in the right hemisphere, whereas their words are dealt with in the left. This is why people who have lost their speech following a left-sided stroke are encouraged to use nursery rhymes to recover speech. These simplest forms of spoken-word song are stored mostly in the undamaged right hemisphere.

Before scientists knew that different neural networks mediated music and language, they knew that some people afflicted with serious speech deficiencies had normal musical abilities. Russian composer Vissarion

Yakovlevich Shebalin composed his first symphony while still a school-boy. He was elected a professor in the Moscow Conservatory where he tutored many well-known Russian composers. In 1953, at the age of fifty-one, he suffered a mild stroke that impaired his right hand, the right side of his face, and disturbed his speech. He recovered, but on October 9, 1959, he suffered a more severe stroke that partially paralyzed his right side and nearly destroyed his ability to speak. He partly recovered but continued to find it difficult to talk and understand what was said to him. He died from a third stroke on May 29, 1963. Remarkably, he completed his fifth symphony only a few months before his death. Dmitry Shostakovich described it as "a brilliant creative work, filled with highest emotions, optimistic and full of life."

Even if we were to accept that music originated from language, the theory does not explain the remarkable elaboration of making music or tell us why making music benefited our ancestors. It also fails to provide an explanation for our strong emotional responses to music.

Origin from the Imitation of Animal Sounds. A theory for the origin of music favored by Valerius Geist, William Benzon, and Bernie Krause, among others, is that music developed from imitations of the vocalizations of other animals.[17] Certain brain regions contain mirror neurons that are active both when we execute an activity and when we observe another individual engaged in that activity.[18] Thus, imitation has a deep neural basis.

Vocal mimicry is attractive as a hypothesis because it doesn't require natural selection to invent something from nothing. A person that imitated an animal's vocalizations might also attempt to imitate its movement and behavior, leading to ritual and dance. Bushmen render typical gaits of animals by softly and rhythmically beating their bows.[19] Animal calls are part of music in many cultures.[20] The Kayabi of the Mato Grosso, Brazil, imitate voices of birds, otters, monkeys, and jaguars. Eskimos imitate cries of geese, swans, and walruses to bring these animals within shooting range.[21] Australian aborigines mimic a hawk alarm call to get a running lizard to freeze so that it becomes an easier target.

Imitating the vocalizations of other members of the same species could serve several purposes. Songs advertise possession of a territory and establish relationships among neighboring individuals. Many birds learn the local dialect, which reduces negative interactions because established individuals recognize one another and avoid pointless contests. Yet, vocal mimicry plays only a minor role among any species of primate other than humans.[22]

Imitation requires voluntary control over the vocal apparatus, a necessary precursor to music. This control in turn may have favored a modified vocal apparatus to facilitate creation of a wide variety of sounds. As Susan Blackmore suggests, "Imitation requires three skills: making decisions about what to imitate, complex transformations from one point of view to another, and the production of matching body actions."[23] It also requires knowing in what context to use different imitations. Our closest primate relatives appear to lack this skill set. Nevertheless, some birds are skilled vocal imitators even though they have much smaller brains and simpler sound-producing structures than primates. Surely primates would have used imitation if it had been advantageous.

Although the ability to mimic other animals was certainly of direct value to our ancestors when hunting and in social interactions, this theory too falls short. We need additional factors to account for the elaboration of music and the remarkable strength of emotional responses to music. Mimicry provides a start, but by itself it cannot yield ecstasy.

Origin from Mother-Infant Communication. Art theorist Ellen Dissanayake has championed the theory that music developed from communication between mothers and their infants.[24] Indeed, the structure of music and lullabies is similar in cultures around the world. Infants and mothers bond strongly using music and body rhythms.[25]

Infant-directed speech—what we call "baby talk"—has a higher overall pitch, a wider range of pitch, longer vowels and pauses, shorter phrases, and more repetition than normal adult speech. Human infants are interested in, and sensitive to, rhythms, tempos, and melodies of speech long before they are able to understand meanings of words. These exaggerated signals are processed as "good" and "affiliative" in the infant's brain, where they are translated into emotions of delight and pleasure. The infant-directed speech and the facial expressions and actions that accompany it reinforce emotional bonds between mother and infant. Eventually cooing evolved into sung infant-directed speech, what we call a lullaby.

Nonetheless, mother-infant bonding falls short of explaining the powerful emotions generated by performing and listening to music. Baby talk probably developed after the human nervous system was already sensitized to melody and rhythm.

Origin from Territorial and Alarm Calls. Gibbons, chimpanzees, gorillas, orangutans, langurs (*Presbytis*), colobus monkeys (*Colobus*), and howler monkeys (*Alouatta*) produce long calls that they use to advertise their claim on territory and broadcast the location of individuals, food, and predators (figure 8.2). These long calls consist of loud, pure tones that accelerate and then slow down toward the end. Long calls are given primarily by dominant males or during duets by mated gibbons. Callers defend their territory and alert close relatives to resources or danger. These situations are important for group members; so they should evoke strong emotional responses in all recipients. To convert them into a simple type of song only a steady rhythm would need to be added.[26] Our early human ancestors probably used similar vocalizations.

This theory meets the criterion of providing direct benefits to both callers and recipients. It also explains why long calls would have evoked

Figure 8.2. Siamang (*Hylobates syndactylus*) adult calling.
The inflated throat sac is a resonating chamber.

strong emotions. In groups with many males, vocal competition for attention and leadership could readily have led to ever more elaboration of calls and, perhaps, the incorporation of rhythm. However, the theory in its current form cannot explain why the long calls of no other primates evolved greater complexity that became music. Nevertheless, this theory provides an important component of a more comprehensive theory of the evolution of music via sexual selection that I will now describe.

Elaboration by Sexual Selection. As we have already seen, Charles Darwin first proposed that vocal communication in general and music in particular developed as a result of sexual selection. Darwin appears to have been right! Females of some species of birds are more strongly attracted to males with larger song repertoires.[27] Song elaboration in humpback whales also is probably also favored by sexual selection.[28]

Why should females pay attention to song repertoires of potential mates? The "developmental stress hypothesis" explains why ability to produce a complex song repertoire is a reliable indicator of a male's quality.[29] Learning and producing songs is a complex cognitive task. It requires a specialized neural system that develops when the individual is growing rapidly and may experience poor nutrition, diseases, and other stresses. At that time, many other physiological and anatomical systems are competing for the limited energy supply.[30] Therefore a male bird that sings vigorously and well is advertising his health and fitness.

The courtship displays of many animals are accompanied by vocalizations. Most of these displays have parts that carry a high cost in time and opportunity to develop and perform, and, therefore, indicate a performer's quality. As much as antlers or bright plumage, vocalizations carry a promise of good genes. What advantages did singing and dancing offer our distant ancestors? Women in early human societies would have benefited by assessing the physical capabilities of the vigorously dancing and calling men and basing their choices of a mate on that assessment. Males would have benefited by watching females dance and from sizing up the dances of rivals and using that information to assess the reliability of other men as partners during hunting and warring. They would also have assessed the likelihood of success if they challenged other individuals for social dominance.

We can see evidence for the sexual selection theory of human music today in the music-making activities of societies all around the modern world. In both the modern world and throughout human history, making music has been a shared activity, carried out mostly by males in small groups. Observing females may respond with music of their own, but only very recently have women in small groups performed on stage.

Men that play the best music and dance the best are more attractive to women.[31] In Kenya, Kikuyu women listen with keen interest; they present a successful flutist with food and drink as a sign of appreciation. In the Trobriand islands, a man with a good singing voice is sure of success with women. As a native put it, "The throat is a long passage like the *wila* (*cunnus*) and the two attract each other. A man who has a beautiful voice will like women very much and they will like him." Many societies have male competitive singing or playing feasts in which winners receive enthusiastic applause and, sometimes, material gifts.

The reproductive benefits that accompany vocal abilities are evident today. Rock stars can have hundreds of times the number of sexual partners as other males. Guitarist Jimi Hendrix died at twenty-seven from an overdose of the drugs he used to stimulate his musical imagination. During his short life he had sexual relations with hundreds of women, maintained parallel long-term relationships with two women, and fathered children in the United States, Germany, and Sweden. Under ancestral conditions, without modern methods of birth control, he likely would have fathered many more.[32]

Of the various factors suggested as explanations for the origin and development of music, sexual selection best accounts for the incredible emotional elaboration of making music. But, if so, why did similar elaboration of vocalizations not occur in other primates? What factors favored the evolution of singing and dancing in our ancestors?

An important factor is that early humans came to live in complex societies with strong competition among individuals for access to resources and mates. Also, the length of time that infants remained dependent greatly increased in the lineage leading up to modern humans, requiring males to commit to provide for the long-term care of their offspring. Women then became sexually receptive throughout the year; the time of ovulation was no longer obvious. These two features of the human female reproductive cycle increased women's options of mate choice and enabled women to choose one male to father her children and a different male to care for them. In combination, these developments would have intensified sexual selection. They are powerful today.

Musical Universals

Music has deep evolutionary roots, so music in all cultures should share basic structural features. Yet, the search for universals in music has a long but relatively unproductive history. Failure prompted some researchers to conclude that music does not share any deep structures, but investigators may have been searching for the wrong things. The striking differences among musical traditions drew researchers' attention to the possible explanations for those differences, distracting them from the search for underlying similarities. Researchers' concentration on differences is why the common deep structure of all languages remained undiscovered until the 1960s.[33]

Technology available in the last quarter century has given us new views of, and new insights into, the musical brain. Positron emission tomography (PET) and functional magnetic resonance imaging (fMRI) studies show that both language and music depend on widely distributed neural networks, some of which overlap. Our complex auditory communications probably evolved concurrently with enlargement of our brain and accompanying cortical lateralization.[34] People who have had fewer than three years of music lessons during their lives recognize tunes more quickly if they are played to the left ear through a set of earphones, and, hence, are monitored by the right hemisphere of the brain, than if played to the right ear.[35] One reason for believing that all music shares basic structural features is that two elements—melody and rhythm—are universal components of music. A third component—harmony—appears to be restricted to certain musical traditions. We'll now look at each of these elements in turn.

Melody. The oldest known pieces of music are pure vocal melodies.[36] Surviving Paleolithic cultures throughout the world use two clear-cut melodic styles. One is a "tumbling strain," a form used by some birds, a wild melody that begins with a leap up to a loud high note. The voice then descends by jumps or glides to low, almost inaudible notes. It then leaps again to the highest note and repeats the descending cascade

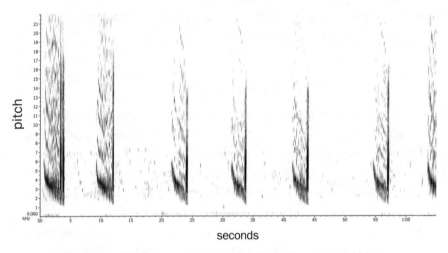

Figure 8.3. A spectrogram of the tumbling song of a canyon wren (*Catherpes mexicanus*). A spectrogram plots the frequency (pitch) of the sound over time.

(figure 8.3). These strains, which are similar to shouts of joy or wails of rage, may derive from such outbursts.[37] Tumbling strains are difficult to execute. Oliver Sachs believes that originally tumbling melodies were fierce and unruly, with no pattern to the steps. Gradually, however, the strains changed to have recurrent patterns.

A second, less emotional type of melody may consist of only two pitches sung in alteration. The voice moves up and down and more or less describes a horizontal zigzag line.[38] These melodies are similar to both the one-note and the two-note patterns of the first babble-songs of small children.

Rhythm. All music has rhythm. We now know much about how the brain generates rhythm. Different sides of the brain process rhythmic perception and harmonic perception. The basic perception of harmonic intervals is seated in the auditory cortex of the right brain. Rhythmic skills favor the left brain.[39] Batteries of tests used to measure musical abilities show that we can be divided into two groups—one has tonality skills, the other has rhythmic skills. Some of us are brilliant harmonically but inept rhythmically, and vice versa. Alas, some of us possess neither skill.

Harmony. Polyphony, the performance of more than one note at a time, also has ancient roots. A rudimentary polyphony with at least some sort of mingling of notes and parts, characterizes some melodies in most cultures.[40] However, all traditional Chinese music is melodic, but not harmonic. Han musicians all play versions of a single melodic line. A few genres of Arabic music are polyphonic, but the vast majority of Arabic music emphasizes melody and rhythm, not harmony.

The highly developed harmony of Western music began with the chants of medieval Christian monks.[41] They had a single melodic line that wavered up and down no more than a step or two. Most notes were held long, and there was no beat (figure 8.4). Chants were prayers; words were much more important than tones.

Harmony depends on the fact that some sounds go well together; that is, they are consonant. Sounds that do not go together are dissonant. Neurology, acoustics, and music theory all contribute to an understanding of dissonance.[42] The neurological explanation resides in the structure of the middle ear. A pure frequency sound stimulates the highest level of activation of receptors at a particular point along the membrane, but receptors on either side of that point also fire. Two frequencies form a dissonant interval when their critical bands overlap and upset

Figure 8.4. The Laborde chansonnier, a medieval songbook owned by the Marquis de Laborde (1807–69). Manuscript on vellum, made in France around 1470.

each other's activations, which is why tones very close in frequency are dissonant.

Ultimately, dissonance is a lack of order; consonance is the presence of order. However, we all know that our ears and brains adjust to dissonance that earlier or untrained listeners found intolerable. Composers may use dissonance to engage listeners by moving away from the tonal center of a composition far enough to cause some anxiety but not so far as to destroy a composition's tonal center.

Anticipation and Musical Pleasure

We listen to music because it evokes strong emotions, most of them pleasurable. Yet, during most of the past several centuries, Western aesthetic philosophers resisted the idea that the arts in general and music in particular have anything to do with "mere pleasure." Music critic Eduard Hanslick (1825–1904) regarded physiological and psychological responses to music as "pathological." Most Western aesthetic philosophers use the word "pleasure" to imply some kind of crude bodily sensation, the study of which was not worth the attention of sophisticated people.[43] But all pleasure is biological. The hypothesis that making and listening to music are not motivated primarily by pleasure is implausible.

Pleasurable Sounds in Nature

If responding positively and negatively to the sounds of nature is one of the roots of human music, the sounds and patterns we find pleasurable in music and nature should be similar. This may be true.

Birds whose songs we find pleasant and musical use rhythms similar to those of our own music. When birds compose songs they often use the same rhythmic variations, pitch relationships, permutations, and combinations of notes as human composers.[44] Singing humpback whales use phrases of a few seconds in duration; they create themes out of several phrases before singing the next theme. Although they are capable of singing over a range of at least seven octaves, humpbacks use musical intervals between their notes that are similar to intervals of human musical scales. Their songs contain repeating refrains that form rhymes.[45] Whales mix percussive or noise elements in their songs; they do so in a ratio similar to that used in Western symphonic music. Some humpback songs include a statement of a theme, a section in which it is elaborated, and then a return to a slightly modified version of the original theme (that is, the A-B-A stanza form).

Many birds elaborate their vocalizations in ways that correspond to a "theme with variation."[46] Young individuals practice singing and develop their musical repertories by listening to themselves and the sounds of others. Some birdsongs use the same rhythmic variations, pitch relationships, permutations, and combinations of notes as human composers. Indeed, birdsongs contain every elementary rhythmic effect found in human music.[47] Humans regard animal calls that use pure tones and rhythms similar to those of our music as pleasant. Pitch intervals in birdsongs that are the same as or similar to the intervals of human music are likely to be ranked high musically. One of the main features of birdsong is its pleasant tonal quality, achieved by production of pure sounds with a restricted frequency range, with few harmonics or overtones. The songs of the musician wren and common potoo, generally regarded as among the most beautiful of birdsongs, use the pentatonic scale, which is also a common feature of traditional music.

Culture and Training Influence Tastes in Music.

Despite its remarkable variety, human music conforms to fundamental patterns in much the same way that a deep grammar underlies all human languages. Nevertheless, although some form of universal musical grammar may exist, the music of other cultures may be so different from the one in which we are raised that we may judge such sounds to be unmusical. In our increasingly global society, this is beginning to change, with an increased audience and market in the West for "world" music with its challenging structures, tonalities, and dissonances.

Within a culture, musical innovation may be unpleasant or even outrageous, when first heard. Often, however, new rhythms, melodies, or harmonies come to be accepted as pleasant after people have listened to them many times. Most people acquire their musical tastes during adolescence among friends of the same age; they often retain those preferences for the rest of their lives.

Even though the forging of social bonds was probably not the factor that first led to the development of music, once we had evolved strong emotional responses to it, music clearly assumed a variety of social functions. Today music synchronizes activities, relieves boredom, reduces tension, stimulates group identity, and communicates political and religious messages. Ritual music and dance trigger brain mechanisms that foster social bonding, suggesting that ritual dancing may have been essential for creating the trust upon which all social interaction depends. Indeed, most people develop their personal musical choices by conforming to the preferences of their peers. Our musical lives are influenced by both the music of the spheres and music of our peers.

9

The First

Sniff

Women believe that they have a better sense of smell than men. They are right! On average, women detect odors at lower concentrations and are better at identifying them than men. Even week-old baby girls turn toward novel odors and spend more time smelling them than baby boys do.

Why are women more sensitive to odors than men? One plausible reason is that during much of our evolutionary history women were the primary gatherers of fruits, vegetables, and small animals. A good sense of smell would have improved their ability to locate fruits, judge their ripeness, assess risks associated with eating them, and evaluate the quality of the many different kinds of potentially edible plants present in the savanna environment.[1] An acute sense of smell would have been less useful to men who were hunting elusive mammals and birds. Another reason is that women, being on average choosier than men, would have made good use of odors in choosing the fathers of their children.

We also use body odors as tribal markers. The Desana people of Colombia believe that each tribe has its own characteristic odor, owing to heredity and what its members eat. The Desana, who are hunters, exude the odor of the game they consume; their neighbors, the Tapuya, who eat primarily fish, are thought to smell of fish. The neighboring farmers, the Tukano, are said to smell of roots and vegetables. Avery Gilbert, an expert in human odor perception, writes that every culture has a foul-smelling food that all members must ingest to be recognized as true natives.[2] You are not really Icelandic unless you eat *hákarl* (rotten shark meat). Real Japanese eat a creosote-smelling mass of fermented soybeans called *natto*. People in northern Sweden make *surströmming* by fermenting herring in barrels for a couple of months. They then put them into tin cans for up to a year. The strong odors released when the can is opened are often compared to rotten eggs, vinegar, and rancid butter. *Kiviak*, a traditional Inuit food from Greenland, is made of hundreds of auks, including their beaks, feet, and feathers, preserved in the hollowed-out body of a seal. Air is removed from the sealskin, which is sewn up and sealed with grease. The foul-smelling fermented birds are eaten during the winter, particularly on birthdays and weddings.

Gilbert may exaggerate, but cuisines do function as strong cultural identity markers. Odoriferous foods are particularly strong markers; they are especially prevalent at high latitudes. True believers claim that they really find them tasty. Although we are poorer than most mammals in detecting chemical signals, odors are clearly important in our lives. To find out why, let's first look at the odors that nature produces.

Nature's Odors

Some of nature's odors are by-products of geologic processes. Volcanoes, fumaroles, and geysers have distinctive sulfurous odors. Lightning produces odorous airborne chemicals. We easily recognize the odor of sea air, but biological processes generate most of nature's odors. Trees emit pungent resin; decaying plant matter gives loam its distinctive earthy scent. Animals mark territories with urine and musk. Ruminants like sheep and goats belch to get rid of the gas created by fermenting leaves in their rumens.

Ripening fruits emit chemicals that attract frugivores from great distances. Plants synthesize many defensive chemicals that protect them from animals that want to eat them. Some of them are emitted only when tissues are damaged,[3] but others are naturally released into the air.

Chemists have identified more than seven hundred flavor-producing volatile chemicals in food and beverages.[4] A single fruit or vegetable may synthesize several hundred, but only a few determine the "flavor fingerprint" that helps animals and humans identify them. As fruits ripen, larger molecules are converted to smaller ones that better nourish seeds. Those same chemicals may attract birds and mammals that eat the pulp and disseminate the seeds. Plants defend themselves against predators—especially insects and mammals—in part by synthesizing toxic chemicals. Herbivores can detoxify many of them, but they must expend energy to do so.

Animals also produce many volatile chemicals. Some, such as body odors, are by-products of normal metabolism. They did not evolve to facilitate communication. A class of chemicals called pheromones, however, evolved in the service of attracting mates and deterring enemies. Animals also use pheromones to mark borders of their territories and foraging trails that other individuals use to find distant food sources. The distance over which these chemicals can be detected and how long they persist in the environment fit their communicative functions.[5] For example, to mark their trails, ants deposit large molecules that diffuse slowly. They spread slowly and persist so that other ants can follow the trail. The sex attractants produced by insects, mammals, and some other animals, on the other hand, are small molecules that disperse widely. They can be detected at great distances by potential mates.

Microorganisms also produce many volatile chemicals as they break down food. The metabolism of bacteria produces foul odors that make rotting food repulsive to animals that would otherwise eat it.[6] In warm climates, vertebrate scavengers must find dead bodies quickly before bacterial products reach toxic levels. In winter, when cold temperatures inhibit bacterial growth, vertebrates can dine on a carcass for many months.

With this background, we are now ready to consider how chemicals inform us about important things in the environment and how we use them to respond to life's challenges.

Odors and Affordance

What odors do we humans pay attention to? We respond negatively to, and avoid contact with, odors we associate with harm or that are harmful in and of themselves. Most volatile chemicals that come from water, caves, volcanoes, and fumaroles make us sick if we inhale them. Fires produce distinctive odors that our ancestors would at first have avoided. Later they learned how to use fire to protect themselves against predators, to cook food, and to create fresh green growth that attracted herbivores. We seek other odors, consciously or unconsciously, because they can help us find a meal or a mate or even communicate with the spirit world.

As most people know when they've tried to enjoy a meal with a bad head cold, the senses of taste and smell are closely linked. If the sense of taste were located in our mouth, simply swallowing a mouthful of wine, would give us the full sensation of its taste. But, the nerves that receive other than the four basic sensations are in the upper nasal cavity where air does not circulate during normal breathing. This is why it is difficult to "taste" food when our nasal passages are clogged. It is also why a food can smell bad but taste good (blue cheese, durian) or smell good but taste bad (coffee).

Odors coming from other living organisms are sources of many kinds of positive and negative information about current or future food sources or enemies. Many organisms produce chemicals that make them less desirable as food. Odors produced by flowers and fruits are designed to be attractive to at least some animals. Many animals emit odors that signal possession of a territory or attract sexual partners. Products of genes in the major histocompatibility complex (MHC) have odors that choosy mates can detect. Female mice prefer as mates males whose MHC alleles differ from their own. Females that mate with MHC-dissimilar males are less likely to abort their fetuses. Humans are not as discriminating as mice, but we also use odors to make mating choices. In 1995 Claus Wedekind and his colleagues found that in a group of female college students who smelled T-shirts worn by male students for two nights (without deodorant, cologne, or scented soaps), by far most women chose shirts worn by men of dissimilar MHCs. Their preference was reversed if they were on oral contraceptives.[7] Several studies since then have shown that we use body odors when choosing mates.

We are attracted to the odors of flowers; many perfumes are derived from them. We locate most flowers visually; we must stick our nose into most of them to detect their odors. Night-blooming flowers emit powerful odors that attract bats and moths from great distances. We can detect those odors, but we seldom approach their source. We have modified flowers to enhance their beauty and increase their life spans. Unintentionally, in the process they have become less fragrant just as the larger fruits we have favored are often less tasty than their smaller progenitors.

A plant will have more surviving offspring if its fruits are not eaten until the enclosed seeds are mature. Immature seeds may pass unharmed through a frugivore's digestive tract, but they are unlikely to germinate. Unfortunately from the point of view of plants, the world is full of animals that attack fruits to consume seeds rather than the surrounding fleshy pulp. They don't wait for the pulp to sweeten. These seed raiders range from large parrots and pigeons to small insects whose larvae live within fruits. Plants defend their unripe fruits with toxic chemicals, but once fruits are ripe, pulp-eating visitors are welcomed. At that time, plants start to produce chemicals that attract animals. Among the chemicals emitted by ripe fruits is alcohol.

Alcohol's Attraction.   Alcohol relieves pain, stops infections, lubricates social exchanges, and contributes to a general sense of joy. Every human society has discovered and taken advantage of these medicinal and psychological benefits of alcohol. People use alcohol to help them communicate with their gods and their ancestors, from the fermented maize beer of the Tarahumara shamans of Mexico to the ceremonial wine of the Jewish and Christian traditions. Cereal grains may have been domesticated primarily to produce beer rather than bread. In an article in the *Scientific American*, Robert Braidwood argued that a single processed food—barley bread—was the driving force for domestication of cereals. Jonathan Sauer responded by arguing that beer rather than bread was the primary motivator. Braidwood then organized a conference titled "Did Man Once Live by Beer Alone." Many opinions were expressed, but no resolution was reached.[8] In any event, probably neither beer nor bread came first because other more easily fermented substances were probably discovered and exploited before barley was domesticated.[9] Barley beer is, in fact, more nutritious than bread. It contains more B vitamins and the essential amino acid lysine. It is also a potent mind-altering and medicinal liquid.

William James vividly expressed the power of alcohol in *The Varieties of Religious Experience*:

> The sway of alcohol over mankind is unquestionably due to its power to stimulate the mythical facilities of human nature, usually crushed to earth by the cold facts and dry conditions of the sober hour. Sobriety diminishes, discriminates, and says no; drunkenness expands, unites, and says yes. It is in fact the great exciter of the Yes function in man. It brings its votary from the chill periphery of things to the radiant core. It makes him for the moment one with truth.[10]

Many other animals are also attracted to alcohol. Fruit-eating vertebrates inevitably ingest alcohol when they eat the sweet mixture of sugar

Figure 9.1. Black spider monkey (*Ateles paniscus paniscus*) eating the fermenting fruit of a Sapotaceae plant, Amazonia, Brazil.

and alcohol that oozes from ripe fruits on branches or on the ground. Some of them have even been observed binging on rotting fruit.[11] In northern Australia, the period just before the rains begin is called the "drunken parrot season." At that time people pick up scores of intoxicated red-collared lorikeets from the streets of Darwin and take them to an animal hospital to recuperate. Orangutans and elephants travel for miles to find fermented fruits. They appear to like getting drunk, but getting drunk in a world full of predators does not seem like a good idea (figure 9.1). What is going on?

The "drunken monkey" hypothesis of Robert Dudley proposes that a strong attraction to the smell and taste of alcohol helps fruit-eating vertebrates find ripe fruit.[12] As a fruit ripens, yeasts on the skin and within the flesh convert sugars into alcohols, primarily ethanol. Ethanol content rises rapidly and may reach concentrations as high as 1 percent. Overripe fruit on the ground may have ethanol concentrations as high as 4 percent, as high as the level in some hard ciders. Ethanol is a small molecule that disperses readily in air. An animal able to detect it could find ripe fruit easily by moving upwind. Finding fruits quickly is beneficial because competition for fruits in nature is often intense. Ripe fruits are often consumed quickly by a variety of birds and mammals and are soon rendered unpalatable by bacteria and fungi in tropical heat.

Although the drunken monkey hypothesis predicts that frugivores should be attracted to the odor of alcohol, when given a choice in the laboratory they typically prefer fruits with less ethanol.[13] This result is still compatible with the drunken monkey hypothesis; frugivores might be attracted to the odor of ethanol and use it to find ripe fruit while still preferring to ingest less of it. As we mentioned, getting drunk in a world full of predators is not a good idea.

In support of Dudley's drunken monkey hypothesis, the benefits of consuming alcohol appear to outweigh the detriments for human health. The good news from an increasing body of research is that moderate consumption of alcohol, particularly red wine, confers health benefits.

Our ancestors would have found it difficult to consume enough naturally fermented fruit to make them sick. However, about ten thousand years ago, humans learned how to control fermentation. Distillation made drinks with much higher alcohol concentrations widely available. Alcohol abuse soon followed. Alcoholism, along with obesity, is a disease caused by great differences between our prehistoric and contemporary environments. We find it hard not to binge on addictive substances that during most of our evolutionary history we could have consumed only in health-promoting moderation.

Body Odors and Perfumes

DON GIOVANNI: *Zitto! mi pare sentir odor di femmina!*
(Hush! I think I scent a woman!)
LEPORELLO: *Cospetto! Che odorato perfetto!*
(Wow! what a nose!)
DON GIOVANNI: *All'aria mi par bella.*
(And a pretty one at that.)

—Mozart, *Don Giovanni*, text by Lorenzo da Ponte

People in all cultures use chemicals to alter the odors their bodies emit. Fragrances have been in use for at least five thousand years; traditional scents are still used in modern perfumes.[14] Manufacturers of perfumes today use the same techniques as their Egyptian predecessors.[15] Many of today's ingredients—cassia, cinnamon, sandalwood, styrax, benzoin, jasmine, and rose—were used as incense by Chinese, Indian, or Egyptian cultures five thousand years ago. Some detailed recipes for myrrh, labdanum, galbanum, and olibanum recorded in the Bible (Exodus 30:34–36) are used today.

Why do we use perfumes? Perfumes can mask unpleasant body odors, but a small number would do that job. Advertising campaigns tell us that the primary purpose of perfumes is to enhance one's sexual attractiveness (figure 9.2). True, but that does not explain why they work. That hypothesis does not tell us why we find those odors attractive rather than repulsive, what messages they send, who receives the messages, or who is manipulating whom.

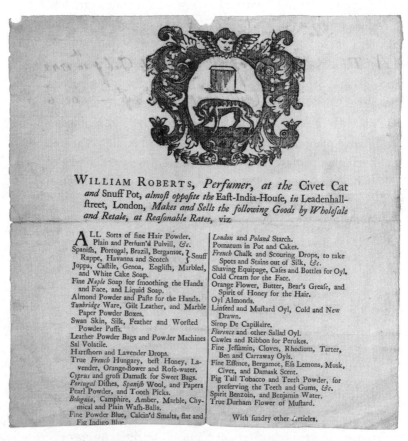

Figure 9.2. Eighteenth-century perfume shop advertisement with civet cat and snuffbox. The musk of the African civet (*Civettictis civetta*) has been prized as a component of perfume for centuries.

An interesting hypothesis proposes that perfumes enhance rather than block a person's natural odor.[16] Why might that work? A plausible answer is that parasites evolve so rapidly that every human generation needs new combinations of genes to resist them. The genes that play that role, the major histocompatibility complex (MHC), are the most polymorphic of vertebrate genes. Determining which MHC alleles potential mates would have been especially important in disease-rich environments.[17] This may be why people typically take a long time to find "their perfume" and then stay with it for many years. The painter of *Les Fleurs Animées* obviously knew that humans prefer other perfumes for themselves than on their sexual partner (figure 9.3). Wise men also avoid choosing perfumes for their partners.

Our Changing Responses to Nature's Odors

We have strong positive and negative responses to nature's odors. We respond positively to odors emanating from objects that afford positive opportunities and negatively to odors coming from objects (rotten food) or situations (a forest fire) that are best avoided. Both positive and negative aesthetic responses to aromas evolved in the service of finding food, avoiding toxins, and selecting mates. We have particularly strong responses to odors that tell us about dangers and about the quality of potential food items. Odors also play a major role in how we interact with one another, advertise our abilities, and integrate ourselves into social groups. Our emotional responses to odors change as we mature. As adults we avoid objects that emit odors of decay, but children do not. Anyone who has raised children knows that two-year-olds insert almost anything in their mouths, including feces, sampling the world while still protected by breast milk. Failure to ingest microbes during infancy, common in today's sterile environments, may be one reason for the current rapid increase in allergies in industrial societies.

A by-product of our use of odors in social interactions is the ease with which our responses to odors can be manipulated by suggestion. A good example is an experiment performed at the University of Wyoming in 1899 by a chemistry professor, Edwin Slosson. He told the class that the experiment's purpose was to demonstrate the diffusion of odor through air. He then poured some liquid from a bottle onto a wad of cotton, keeping it far

Figure 9.3. Tubereuse and jonquille from *Les Fleurs Animées*. The man's flowers are orange, the woman's are white, in the colored original.

from his nose. He started a stopwatch and told the students to raise a hand as soon as they smelled something. He reported the results as follows:

> While awaiting results I explained that I was quite sure that no one in the audience had ever smelled the chemical compound which I poured out, and expressed the hope that, while they might find the odor strong and peculiar, it would not be too disagreeable to anyone. In fifteen seconds most of those in the front row had raised their hands, and in forty seconds the "odor" had spread to the back of the hall, keeping a pretty regular "wave front" as it passed on. About three-fourths of the audience claimed to perceive the smell, the obstinate minority including more men than the

average of the whole. More would probably have succumbed to the suggestion, but at the end of a minute I was obliged to stop the experiment, for some in the front seats were being unpleasantly affected and were about to leave the room.[18]

Yet, Slosson was holding a cotton ball soaked only in water! Other experimenters have obtained similar results.[19] The members of the "obstinate minority" must have overcome strong urges to raise their hands, just as we are motivated to laugh at a joke even if we failed to "get" the punch line. The desire to be an integrated member of a group often trumps reality.

The underlying reason for the power of "spin" in relation to odor perception is that smells don't happen just in the nose. The brain is active at all times during odor perception, controlling the intensity of sniffing, influencing habituation to current inputs, and preparing us for action. The structures of molecules are not reliable guides to how it smells to us. Molecules exist in the air, but we can register only some of them as smells. "Odors are perceptions, not things in the world. The fact that a molecule of phenylethyl alcohol smells like rose is a function of our brain, not a property of the molecule."[20]

Our response to odors is a dramatic illustration that our nervous system is organized functionally to help us make good decisions, not to provide us with accurate pictures of nature. We still don't understand why we differ so much in our ability to detect odors. Women may, on average, be better than men in detecting odors, but sensory abilities of the sexes overlap. Moreover, great variability exists within each sex, from people born without a sense of smell to "supernoses" who excel in the industries of perfumery, viniculture, and the creation of artificial flavorings. Is this variability the result of differences in individual experiences, such as positive or traumatic events accompanied by strong odors that left indelible impressions? Has there been selection that favored specialization, or is the variability simply a result of sexual recombination? Keep inhaling!

10

Ordering
Nature

Ornithologist Frank M. Chapman was one of many people concerned about declining bird populations more than one hundred years ago. His worries led him to propose a new holiday tradition—a Christmas bird census. Rather than hunt birds, he suggested that people go out and count them. He inspired twenty-seven birders in twenty-five cities in North America to conduct the first Christmas Bird Counts on Christmas Day 1900. The idea caught on. In 2011, birders held Christmas Bird Counts in 2,215 different places around the world; they tallied 61,359,451 individuals (figure 10.1). Communities compete to tally the most species or to get the highest number of individuals of particular species. For some people, participating in the counts has become an obsession; they rise before dawn to join in several counts within a few days; counts can now be held over a two-week period.

Why are so many people motivated to spend long hours, often in cold, miserable weather, counting birds? Why do they care how many different kinds they see? Why do they not just count the number of individuals? Why do they care who sees the most species? A plausible evolutionary reason is that an environment rich in species offers greater affordance than one with few species. The resources our ancestors required (food, fiber, and shelter, among other things) came from many species. Knowledge of those species would have helped them find them. When our ancestors evaluated habitats, they surely thought about the species living there and how they could use them. Paying attention to other species evolved to become pleasurable. Later we'll see how this ability to distinguish among and remember many different kinds and forms of things expresses itself in ways that don't always seem directly connected to life on the African plains.

Figure 10.1. Jeannie Elias, *left*, Mary Spencer, and Alison Wagner look for birds in Fayston, Vermont, December 19, 2008, as they take part in the National Audubon Society's annual Christmas Bird Count. Between December 14 and January 5, birders around the Western hemisphere headed out on one day to keep track of the birds they could see and hear in a fifteen-mile diameter circle.

Counting Species on the Savanna

Throughout human history, knowledge of other species, especially their suitability as food, has been crucial to our survival. Paying attention to timing and location of flowering plants would have told our ancestors where they could find fruit in season. Signs of recent animal activity—tracks, broken branches, scat, and odors—provided hunters with valuable information. Movements of herds of mammals and flocks of birds offered direct information about food availability. Humans have long observed other species of animals to determine what is safe to eat.[1] The Incas learned which plants were dangerous to eat by observing plants eaten by animals that later became sick.[2] The behavior of female mammals might indicate where their vulnerable offspring were hidden.

Animals could be valuable in other ways. By observing them, early humans could benefit from animals' knowledge of the environment. Seafarers have long used the behavior of seabirds to aid navigation.

People of the Poluwat Atoll in the Caroline Islands navigate across great distances between islands by dead reckoning, but when they become lost or stray from their intended course, they rely on behavior of seabirds to find land.[3] The islanders have intimate knowledge of the birds' foraging behavior and how their flight destinations change between dawn and dusk.

Before the invention of agriculture people typically occupied a series of seasonal campsites, following seasonal availability of animals and plants. People had to decide when to shift from a dry-season camp to a wet-season camp, or from a winter camp to a spring camp. These decisions to shift camp needed to be timed carefully, and this depended on a deep understanding of life cycles of local species. This deep understanding of other species lies at the basis of our attraction to the great diversity of living things. I argue that it's also the basis of a wide range of human collecting behavior.

Aesthetic Responses to Biodiversity

Although environments that are species-rich offer greater affordance than ones with fewer species, the relationship is not simple. Adding a few species to a species-poor environment may greatly increase its value, but a few more species in an already species-rich environment may have little effect. Also, the more different species there are, the more difficult it becomes to track them. Research shows that we prefer environments with intermediate levels of biological richness to both simpler and more complex ones.[4] Species-poor environments have too few resources; species-rich ones have so many that choosing among them or remembering where they are becomes difficult.

People in traditional societies typically profess a holistic view of nature in which each species is perceived to have a specific function. This often extends to a belief that nonhuman species are important for spiritual reasons beyond their utility as food, fiber, other materials, or medicines.[5] Many people in modern industrialized societies believe that a world with many fewer species would be a less desirable place to live in. This belief spurs them to donate to organizations that work to preserve species that live on other continents, places they may never visit.

Knowing that those species continue to exist is sufficient motivation for donating.

Although people are attracted to a variety of organisms, we find some environments with few species very attractive, if we can judge by the kinds of environments we recreate in our gardens and other highly humanized landscapes. The most highly developed garden traditions—European formal gardens and Japanese gardens—are dominated by a few species of woody plants. Most of us respond positively to gardens that display a profusion of flowers of different species and colors, but environments that contain a jumble of plants of many species receive low scores in psychological tests (chapter 6). Subjects report that they are too difficult to interpret, that it is hard to determine how to enter and use those environments. We respond positively to large flocks of birds and herds of mammals that have only one species, but I am not aware of experiments that test our responses to scenes of flocks and herds that differ only in the number of species in them. Opportunities for imaginative research abound.

Classifying Species

To know how to respond to different species it helps to categorize them. By doing so we create stereotypes that greatly simplify deciding what to do with their members. By the logic we have employed throughout this book, classifying things should be pleasurable. It is! As psychologist Nicholas Humphrey pointed out, pleasure happens when we view and attempt to order shapes and patterns because "an activity as vital as classification was bound to evolve to be a source of pleasure."[6]

Naming things, particularly plants and animals, was probably a major function of early human language. We know that classifying organisms was an important activity in ancient Mesopotamia and the Mediterranean basin.[7] We can classify things that lack names, but it is much easier to remember things if they have names. Ancient Hebrews recognized the importance of names by having Adam, as his first task, name the animals. "So out of the ground the Lord God formed every beast of the field and every bird of the air, and brought them to the man to see what he would call them; and whatever the man called every living creature, that was its name. The man gave names to all cattle, and to the birds of the air, and to every beast of the field" (Genesis 2:18). The writers of Genesis had poor knowledge of the extent of Earth's biodiversity; they imagined Adam's job to be a fairly simple one that he could accomplish in less than a day, and they did not care if plants had names.

We now classify anything that can be ordered, but the value of paying attention to other species may be the origin of our desire to classify things. The pleasure we gain from finding order in nature may also help explain our propensity to seek and find "order" where there is none. People find forms of living organisms in clouds. We imagine outlines of people, animals, and human artifacts in rock formations that are the result of normal erosion. We find monsters in driftwood and wave patterns.

Rarity and Affordance

Common species probably provided most of the resources used by our ancestors; they do so in hunting and gathering societies today, but rare and unusual species and events may have provided valuable information about environmental changes. Not all novel events signal something important, but some of them do. Rare events may indicate that current ways of using the environment should be altered. Indeed, unusual events (more powerful hurricanes and tornadoes, early flowering of plants, early breeding of birds) are currently telling us about the consequences of climate change. Rare species have provided special flavors (spices), scarce nutrients, and medicinal benefits (recall chapter 7).

The survival value of attending to rare and unusual events may help explain the aesthetic attraction of unusual variants of plants and animals, an attraction that has led to a proliferation of cultivars of domesticated plants and breeds of domesticated animals. In 1800, only fifteen breeds of dogs were recognized; a century later, there were sixty (figure 10.2). A similar proliferation occurred with virtually all forms of livestock, domestic fowl, pigeons, canaries, and plants. By the 1850s, 150 breeds of pigeons already were recognized.

Animals with atypical coloration have always fascinated people. Museums are stuffed with them, deluding people into thinking that they

Figure 10.2. Prize dogs at the Birmingham dog show, 1869. The National Dog Show Birmingham remains the world's oldest ongoing dog show. The American Kennel Club now recognizes 161 distinct breeds, from the saluki, with its origins in ancient Egypt, to newly recognized breeds like the Norwegian lundehund, trained to hunt puffins.

are much more common than they really are. Albinos or partial albinos are the most common color aberrations. These variants typically arise as random mutations; most of them quickly die out in nature. However, they are nurtured in captivity and incorporated into popular varieties.

Ordering Becomes Collecting

Our remote ancestors could not afford to collect many things because they had to carry their possessions with them when they changed seasonal camps. They probably collected some small things like arrowheads—different kinds are used for killing different types of prey (figure 10.3)—but most things that accumulated at a campsite were probably left behind when people departed. Nevertheless, the pleasure of classifying things has been expanded into a desire to collect them. Our ancestors probably began to collect things as soon as they established permanent villages.

Figure 10.3. A boy displays his prize arrowhead collection
at a hobby show at Hanford, Washington.

Most of us now collect something at some time during our lives. My collecting passion was stamps; I organized mine by country and date of issue. People around the world collect art objects, books, coins, stamps, antiques, and a mind-boggling array of items from autographs to Tibetan mountain ghost traps. Collectors of specific things form organizations and have annual meetings at which they talk about their collections as well as buy and sell. They spend a lot of time ordering them.

We are not the only animals that collect things. Many species of animals cache food for later use. Hoarding, storing, or caching food is known in at least twelve families of birds, twenty-one families of mammals (but not in primates), and many insects.[8] The honey we value so highly is food bees have stored for the savanna dry season or the northern winter. We use the name of animals that collect things to describe people who cannot throw things away—pack rats.

Collecting things sometimes becomes a passion with undesirable consequences. Physiologist Ivan Pavlov, who thought deeply our passion for collecting, expressed it this way:

If we consider collecting in all its variations, it is impossible not to be struck with the fact that on account of this passion there are accumulated often completely trivial and worthless things, which represent absolutely no value from any point of view other than the gratification of the propensity to collect. Notwithstanding the worthlessness of the goal, every one is aware of the energy, the occasional unlimited self-sacrifice, with which the collector serves his purpose. He may become a laughing-stock, a butt of ridicule, a criminal; he may suppress his fundamental needs, all for the sake of his collection.[9]

What induces a person to become obsessed with collecting things? The answer, of course, lies in the brain. A detailed study of eighty-six patients with brain lesions showed that the thirteen patients with lesions in a particular prefrontal region of the cortex developed compulsive, obsessive collecting behavior.[10] They all massively accumulated useless objects and persisted in doing so despite vigorous efforts by their close associates to stop it. The data suggest that a mechanism in that part of the forebrain normally adjusts our predisposition to acquire things so that it does not become disruptive. Not every person who becomes obsessed with collecting things has brain lesions, but something may have disrupted the normal control system.

Evidently our desire to collect things is so strong that a special mechanism evolved to keep us from going overboard. A control mechanism may have been necessary because, as we will see, collections are excellent for showing off.

Sexual Selection and Elaboration

In addition to the enjoyment they get from classifying objects, people take pride in the sizes of their collections. They also take pride in the number of places they have visited or things they have seen. During the eighteenth century, gardeners in Europe eagerly sought exotic plants from around the world and boasted in the number of species they raised. A vigorous trade existed between England and North America. An amateur gardener in Pennsylvania, John Bartram, became the principal supplier of seeds and plants to growers in England.[11]

Many people take great pleasure in finding as many species of birds as they can on a given day or year. They assemble "life lists" of species they have seen. Some people travel hundreds or even thousands of miles to add a rare species to their life list. Why do we engage in such costly and seemingly useless behavior?

Competition for status, often unconscious, may underlie such behavior. We use other species to compete for status, indirectly by flaunting the length of our lists, and more directly by competing in dog shows, horse shows, flower shows, and bonsai shows. People work mightily to win prizes in those shows and proudly display their winning ribbons and plaques.

Catching and keeping birds has long contributed to status. Men gave birds to women as courtship gifts. In medieval times catching small birds was, along with hunting, falconry, and fishing, one of four noble and status-enhancing pursuits. Hunters had higher status than bird catchers, but the latter ranked higher than fishermen. The more successful a bird catcher, the greater was his status in the local community. In his little book *Hungers Prevention*, Gervase Markham claimed that small birds have two uses: "either pleasure or food, pleasure because every one of them naturally, have excellent Field-Notes, and may therefore be kept in cages and nourisht in their own tunes, or also trayned to any other notes, or else for food, being of pleasant taste, and exceeding much nourishing, by reason of their Naturelle heat, and light digestion."[12]

Canaries reached central Europe via Spain and Italy in the 1400s. Their melodious songs immediately captured the attention of people on a continent already full of bird keepers. Germans enlarged the vocal repertories of captive-reared birds by exposing them to large numbers of songs. Several villages became involved with breeding canaries, but the center of the business was St. Andreasberg. During the 1820s, the town was producing about four thousand male canaries each year (only male canaries sing). By 1882, three-quarters of the eight hundred families in St. Andreasberg were rearing canaries. In some years the town exported as many as twelve thousand male canaries. The trade in caged songbirds was not confined to Europe. More than ten million birds were sent to America during the first four decades of the twentieth century.[13]

It was perhaps inevitable that, once we began breeding birds for their songs, people would stage singing contests (figure 10.4). The earliest known formal song contest was a chaffinch competition in the Harz Mountains of Germany in 1456. The species used in early song contests

Figure 10.4. The "Kota Tua Cup," a songbird competition in front of
Fatahillah Museum in the Kota Tua area of Jakarta, Indonesia, January 24,
2009. Hundreds of birds competed in this singing competition.

included skylarks, goldfinches, linnets, greenfinches, and chaffinches.
The winner of a "strong singing" contest was the bird that sang the highest number of complete songs in a set period of time, usually about five
minutes. Success in another type, "distance singing," was judged on the
number of songs a chaffinch performed in thirty minutes or an hour.

When canaries became available, they were also extensively used in singing contests. A winning canary conferred great prestige on its owner. Chaffinch singing contests have fallen out of fashion with the rise of the modern animal-rights movement.

Competition for status, and the sexual opportunities it affords, is not a uniquely human attribute. Other species use attractive objects for the same purpose. Male bowerbirds decorate their elaborate bowers with colored objects that they place near the entrances to their bowers (figure 10.5). They compete for objects and steal them from their neighbors. The number of objects with rare and unusual colors shows a male's status. Only dominant males can steal rare objects from other males and prevent them from recovering them. Objects having colors that are common on the forest floor tell nothing about the quality of their owner because subordinate males also can assemble them.

Collecting organisms and other things has made great contributions to scientific knowledge. Without collections we would not know about the great diversity of life. We know that exposure to DDT caused the shells of birds' eggs to thin because we could measure eggs collected

Figure 10.5. Male great bowerbird (*Chlamydera nuchalis*) at his bower decorated with green glass, plastic toy elephant, toy soldier, and other decorations, Townsville, Queensland, Australia.

before DDT was made. We know how natural environments sounded because Bernie Krause recorded them.[14] But our desire to possess exotic species or parts of them (skins, horns, tusks) is threatening the existence of orchids, parrots, tigers, leopards, elephants, rhinoceroses, and many other species. Most nations of the world have joined CITES (the Convention on International Trade in Endangered Species of Wild Fauna and Flora), established in 1975, to control the massive worldwide traffic in rare species, but animals and animal parts command such high prices in the world market that smuggling continues. We need to learn how to use our drive to collect and categorize things in a way that serves our own desires without endangering other species that share our small planet with us.

11

The Honeyguide and the Snake

Embracing Our Ecological Minds

In the opening episode, a small bird, appropriately called a honeyguide, led Boran hunters in Kenya to a bee-hive. Honeyguides are good at finding hives, but they are unable to open them. The Boran and other tribes have trouble locating widely dispersed hives but are adept at extracting the nutritional bonanza of honey and bee larvae, some of which they leave as payback for the helpful birds. This remarkable partnership evolved because honey was uncommon on the African savanna and both partners craved it. Except during the brief season when fruits were ripe, it was the only source of sugar. Centuries later, as our ancestors spread abound the globe, they clung to their taste for honey and produced a fermented version, mead, the intoxicant that became the libation of Viking sagas and medieval poetry.

A sweet tooth was adaptive for our hunter-gatherer ancestors who needed to store energy to be active during times of food scarcity (the human body converts excess sugar to fat) and to pay for their large, expensive brains. Most of the foods they collected were not sweet. Much later, agriculture gave us abundant starch. Only recently has technology inundated us with sugar. Our savanna-adapted bodies still crave sugar, but lean times rarely or never come for most people in modern technological societies. So now we deposit the excess as fat and become obese. An evolutionary perspective helps explain why about two-thirds of American adults and 10 to 15 percent of children less than nineteen years of age are overweight or obese and why there is a global epidemic of obesity and type 2 diabetes in both adults and, for the first time, children.

Coren Apicella, a biological anthropologist from Harvard Medical School, and her colleagues used our fondness for honey to find out whether social networks in modern societies are ancient patterns or recent developments. They gave members of the Hadza tribe, hunter-gathers in Tanzania, honey sticks (hollow wax straws filled with honey, such as are sometimes sold at drugstores and farm stands). They asked the participants with whom they would share the honey. The Hadza sharing pattern closely resembles the sharing behavior of people in the developed world.[1]

As the Boran hunters followed honeyguides flying through the trees, they would have been advised to look down as well to avoid coming close to a python or adder without seeing it. Their sensitive peripheral vision would have helped them detect motionless snakes in the grass. If one of the hunters had detected a snake, he would have alerted the others, allowing them to give the hidden snake a wide berth. In that way they would have continued to be predators rather than becoming prey.

Our fondness for sweets, honed on savanna honey, important though it is today, is only one of many ghosts of past environments that persist in the modern psyche. Our honey fixation illustrates the central message of this book: the experiences of our distant ancestors in their savanna homeland color our responses to everything around us—our reactions to landscapes, other organisms, and the sounds and odors of nature. Using evolutionary insights we have uncovered legacies born on the African savanna that would have gone unnoticed without an evolutionary perspective.

Savanna Legacies

Some of the responses I have discussed were already well known, but evolutionary psychology has shed new light on why we have them. Consider, for example, our attraction to water. Evolutionary psychologists have shown that infants and toddlers on their hands and knees mouth flat, glossy objects as if drinking, suggesting that our positive response to water is innate. Some of the other responses that we better understand as a result of research by evolutionary psychologists are that we have

- Hardwired autonomic nervous system circuits that respond more strongly to stimuli that were relevant in the past (snakes, dangerous herbivores, and large carnivorous mammals) than to stimuli that pose greater threats today (firearms, electrical wires, speeding cars).
- Strong responses to pointed forms that resemble of the fangs, claws, and horns of dangerous animals.
- Automatic and unconscious association of sickness with food eaten hours earlier and negative responses to likely sources of contagion, such as rotting flesh, and signs of infection, such as pus. This implies an intuitive microbiology that evolved long before we were aware of the existence of microorganisms or had a germ theory of disease.

Although evolutionary psychology has enriched our understanding of those things we already knew about, its greatest value has been to help us think of responses and patterns we would have missed completely, or failed to have imagined, without an evolutionary perspective. Among the surprises that have been discovered are that

- Women remember better than men the location of high-quality food in a supermarket.
- We have a previously unknown and unsuspected vertical illusion; we dramatically overestimate vertical irregularities on the ground.
- Prospect-refuge theory can explain our unconscious emotional responses to unfamiliar environments.
- The savanna hypothesis predicts new explanations for our preferences among tree shapes, and the way we manipulate trees in parks and gardens.
- We remember animal tracks better than most other types of objects.
- We are especially attracted to changing colors of leaves and to flowers.
- Our neural systems more accurately detects changes in the positions of animals than of objects, even cars and other motor vehicles, by far the most deadly moving things in today's environment.
- Children have an intuitive understanding that animals are self-propelled whereas other things are not.
- We have special neural mechanisms attuned to tessellated patterns that facilitate detection of snakes.

Some of the most striking insights evolutionary psychology has provided into human nature involve our responses to other people, an arena

I have only lightly touched on in this book. Among the most important findings are that

- Our brains possess a special neural module for detecting cheaters.
- Our surprising propensity to share what we have with others in "one-shot" economic games may a by-product of social decision making under conditions of uncertainty.

Regarding our propensity to share, our ancestors could not have known whether an encounter with a stranger would be the only one or the first of many interactions. It wouldn't pay to share resources with someone you might never see again, but it could be worse to refuse to share with someone who might end up being a neighbor. No stranger ever came from far away. Under those conditions, evolutionary psychology predicts we should hedge our bets. We should evolve responses that balance the costs of mistaking a one-shot encounter with a stranger for the first of many encounters with someone who will become familiar. Wrongly assuming we will have no future encounters may result in forfeiting the benefits from future social encounters when kindness and generosity are repaid. Even though subjects are told that the experiment is a one-shot encounter, our savanna minds respond otherwise.[2]

The Road Ahead

The body of information on our emotional responses to the environment is impressive and growing rapidly, but it represents only the beginning of what the sciences of behavioral ecology and evolutionary psychology can and will discover. We cannot imagine today the rich array of hypotheses future investigators will develop or what their experiments will reveal.

As Joseph Heinrich at the University of British Columbia has pointed out, most laboratory studies that test evolutionary psychology hypotheses have been carried out with WEIRD (Western, educated, industrialized, rich, and democratic) college students who are unrepresentative of humanity as a whole. Much more cross-cultural research and studies of people of all ages will be needed to determine which responses are universal and which are culture specific, at which ages responses first develop, and which are more easily altered by experience. Some responses that embody strongly held cultural values will probably vary widely, but many other responses may be similar across cultures. The many known "human universals" suggest that, despite the overlay of nationality and culture, we are fundamentally alike.[3] People's preferences for landscape features, especially savanna-like vegetation, tree shapes, and colors, for example, are similar across many cultures. Yet, as scientists generate new hypotheses that predict response patterns we have not yet imagined, we are likely to be in for some interesting surprises. The unexpected outcomes will also be sources of much of the pleasure that will accompany future discoveries. Finding something unexpected is much more exciting than affirming something we judge likely to be true.

Our Complex Ecological Minds

Many of our rich and varied emotional responses to the physical and biological environment were molded when our ancestors lived on African savannas. Their conscious and unconscious assessments influenced decisions about where to go and what to do there. To better understand our complex modern minds, we need to briefly dip back into the past for some ecological context. Rarely did our savanna-dwelling ancestors need to choose among several options at once. Most choices were "do or don't," a choice between responding to a particular thing or situation or moving on to the next decision.

Consider an individual (we'll say a male early ancestor) entering an environment for the first time. On encountering this new place, he decides either to stop and explore or move on and continue searching. To make an appropriate decision he needs to have a guess about the general state of the environment. He must have an expectation of what he might encounter and how soon if he continued to explore. The decision to give chase is based on how long he thinks it might be before another potential meal comes along. Like other foragers, he would have encountered most of his potential prey one at a time. As do starlings, which like most birds encounter prey individually, he would have made better choices by having more information about the general state of the environment.[4]

Even though our ancestors sometimes did need to evaluate and choose among simultaneous options, they always had incomplete knowledge; the consequences of their decisions were always uncertain. Rarely were there "right" answers they could have determined by taking time to gather more information although they clearly benefited by being able to rapidly process much information.

This fundamental feature of the "decision environment" of our ancestors may help explain our behavior today. To see how, let's look at the surprising results of some of the imaginative experiments that were summarized and synthesized by psychologist Daniel Kahneman.[5]

We start with how we size up an offer of easy money. In some experiments subjects are asked to choose between two simultaneous options for which there is a mathematically correct answer. For example, people were asked to choose between receiving $46 (guaranteed) or tossing a coin and getting $100 (heads) or nothing (tails). Most people preferred the sure thing, even though its expected value over repeated trials ($46) is less than the expected value of the coin toss ($50). The reason is that we give greater psychological weight to losses and potential losses than to gains. As we have seen, our ancestors benefited by developing a powerful aversion to loss.

Our negativity bias also shows up in the following type of experiment. People in one group are given a coffee mug; people in another group are given a chocolate bar. Soon thereafter individual members of both groups are told that they can keep the gift they have but if they wish they can exchange it for the alternative. Most people in both groups elect to keep what they have rather than trade. For our ancestors, possession of an object normally meant that they had expended some effort to acquire it, even if, as in many psychological experiments, it was a gift. Moreover, a gift normally implies indebtedness to someone. Therefore, the loss of a possession psychologically implies a significant sunk cost. Our ancestors rarely received something for nothing.

Our ecological minds can also explain the striking binary structure of human thought. Most decisions of our ancestors had to make were "either-or." An object or landscape was either to be approached or avoided. An item was either edible or inedible. An individual was a suitable mate or he or she wasn't. And they often had to make decisions quickly. An individual that pondered the desirability of fleeing from a hungry predator was more likely to be captured than one that fled instantly. A sexually active male that debated at length the desirability of mating with a receptive female would find another's genes in his place.

Although taking action requires quick decisions with polar opposite outcomes, inputs from the environment tend to be complex and often contradictory. To help us decide quickly, our brains resolve complex inputs into polar categories. Polar thinking is not simply a feature of Western thought, a legacy of Greek and Roman logic and languages that gave us polar prefixes: *con-dis, post-ante, pro-con*. Polar opposites are found across cultures, among people speaking different languages and with different intellectual histories. Cosmological schemata of cultures as diverse as Chinese (for example, *yin* and *yang*), Indonesian, Keresan Pueblo Indians, and Oglala Sioux all divide nature into a few categories.[6] Claude Lévi-Strauss, the leading anthropologist of his generation, claimed that all tribal myths are built upon polar opposites—hot and cold, night and day, raw and cooked, good and bad.[7] Spatial distinctions in most languages are relatively simple. Although space is a continuous variable, in our languages something is either close or far.[8] These universal patterns suggest that polar typology has deep evolutionary roots. Our ecological mind explains why we evolved to resolve complex inputs into binary outputs. Action is polar!

Our Ecological Mind and the Social Sciences

The view of the human mind that emerges from investigations of our interactions with the environment contrasts strikingly with the human mind in the standard social science model that prevailed for most of the twentieth century.[9] According to that model, the evolution of human consciousness, combined with our remarkable learning ability, emancipated us from the control of our genome. There is no such thing as human nature. As expressed by José Ortega y Gasset, "Man has no nature: what he has is his history." In other words, the human mind begins as a blank slate. The formless mind of an infant is transformed into a cultural human adult mind by the surrounding culture. Psychologists who adopted this perspective believed that all thought, feelings, and behaviors could be explained by a few mechanisms of learning.

The amazing recent advances in cognitive neurology have made that model untenable. Yet, many people who recognize that our behavior is a product of complex interactions between heredity and environment during maturation still believe that we differ fundamentally from other animals in that environmental influences so completely trump genetic influences on our behavioral development that a potential role for genes on much of our behavior can be ignored. However, some traits, such as the ability to learn a language early in life, appear entirely genetically determined; barring significant developmental delay or a certain rare mutation, any child will speak by the age of two. Other traits, such as which language a person speaks, are entirely environmentally determined. The insights into our ecologically molded minds that I have illustrated in this book greatly enrich our understanding of *which* components of our behavior bear the imprint of our long evolutionary history and *why* they do.

Our Ecological Minds and Sexual Selection

We are constantly sizing up the competition, comparing ourselves to rivals, especially sexual rivals. We make many decisions by estimating our ability relative to that of competitors or opponents. If a male judges the other guy to be larger or stronger, he usually backs down. We rarely pick fights we are certain to lose. Tall men are more successful in business than short men; they typically win political contests. A review of the extensive literature on selection acting on traits that influence mating success, such as elaborate displays in males, shows that sexual selection is stronger than natural selection acting on traits that directly influence survival or fecundity.[10]

A few examples will explain what I mean. Competition for status, which drives many components of human behavior, begins in childhood and continues during adulthood. We cultivate flowers, fruits, and vegetables that are much larger than natural ones. We build homes much larger than necessary to satisfy the basic amenities a house provides. Such conspicuous consumption is a sexual display like a lion's mane or a peacock's tail. An evolutionary perspective explains why we elaborate only a small set of objects and behaviors that are typically things that are intrinsically pleasurable. The psychological effect of elaboration is to make them even more pleasurable. So we build elaborate houses, maintain large, fancy gardens, compete for status with our collections of things, and display the physical skills that would have served our ancestors well. Many dress styles enlarge or exaggerate breasts and buttocks (women) or shoulders and upper-arm strength (men). Competition for status is successful only if it uses objects that generate strong positive emotions. Otherwise, nobody cares.

Sexual selection can also explain why our gardens and parks and other green spaces don't always display the features of a savanna. We often alter them to display our power, wealth, and status. As we discussed in chapter 5, formal European Renaissance gardens differ strikingly from tropical savannas. Many of them are best viewed from the owner's home, not by walking through them (figure 11.1). The elaborately manicured landscapes function as a dominance display. Only wealthy, powerful people can afford to maintain them.

Not all our fascination with elaboration is due to sexual selection. We are fascinated by supernormal scenery—big storms and rugged mountains. Painters and photographers exaggerated the topography of the American West as part of a campaign to convince Congress to establish national parks in the most spectacular areas. We are not alone in thinking that bigger is better. For example, adult songbirds respond to the

Figure 11.1. Garden Castello Balduino from the balcony, Montalto di Pavia, Italy.

Figure 11.2. These gullible reed warblers (*Acrocephalus scirpaceus*) had their work cut out for them when they were tricked into thinking this enormous cuckoo chick was their own.

gaping mouths of their offspring by bringing food. The gaping mouths in a nest are virtually certain to be those of its offspring. Thus, when a cuckoo, cowbird, or other brood-parasitic bird lays her egg in the nest of a smaller bird, the large egg is incubated. More surprising, the parents eagerly feed the exotic nestling, even when it grows to be much larger than their own offspring (figure 11.2).

Applied Environmental Aesthetics

The environmental movement has been concerned primarily with the effects we have *on environments* rather than effects *environments have on us.* We do, of course, have profound influences on Earth's climate and the cycles of carbon, nitrogen, and phosphorus. We are massively modifying Earth's landscapes, moving disease organisms around the world, and causing the extinction of many species of plants and animals. We live on an increasingly cultivated planet with shrinking wildlands like our ancestors knew in East Africa. Our strong positive and negative emotional responses to living organisms and natural environments influence how we respond to nature, how we attempt to manipulate it, and why we care about it. Understanding the roots of our emotional responses to environments, in addition to its intrinsic interest, helps us understand why we manipulate nature as we do and suggests how to improve our role as stewards.

Insights from evolutionary psychology suggest ways to interact with our environment that are both emotionally satisfying and less destructive than much of our current behavior is. We can use our emotional responses to events and situations to design educational programs and environmental policies, and to use plants and animals creatively in a variety of therapeutic situations. There are already some promising applications of environmental psychological research to current problems and challenges.

The Environment and Cognition

To respond appropriately to environmental challenges, our ancestors had to take in a flood of diverse information, compare it to previously learned information, and decide how to act in a new situation. Functioning in a complicated and uncertain environment favored complex cognitive capabilities. One legacy of this interaction between our brains and complex natural environments is that being in natural surroundings actually helps us think more clearly. People suffering from attention fatigue recover when exposed to natural settings, in part because natural places like woodlands and meadows and riverbanks engage the mind effortlessly and provide a respite from sustained attention.[11] Exposure to the great outdoors has been shown to help children with attention deficit/hyperactivity disorder (ADHD), particularly those for which medication has proven ineffective. Children with ADHD concentrated better after a twenty-minute walk in a park than in two other urban settings.[12]

The good feelings we experience when we're out in nature help us re-call information from memory and find creative solutions to problems. Exposure to nature promotes recovery from mental fatigue stemming from demanding work tasks that require a high degree of sustained attention.[13] Stress reduces our performance on various higher-order cognitive tasks, but even a brief exposure to nature leads to improved mood, reduces stress, and results in better performance on experimen-tal tasks.[14] This is true whether subjects are exposed to actual natural settings or simply shown photographs.[15] That exposure to the natural world reduces stress has important implications for the design of work spaces, living quarters, and healthcare facilities. Architects and plan-ners are just beginning to incorporate these insights into their blue-prints.[16] Evolutionary psychology results would predict that exposure to savanna-like natural environments will yield better results than expo-sures to other landscape types, but this prediction has not been tested.

Aesthetics and Environmental Education

As we have seen, emotions are central to human decision making; we communicate emotion in part through the arts. This suggests that educa-tors should make greater use of the performing arts to inform audiences about conservation issues and to stimulate dialogue and inspire action. Conservation education and outreach at all levels can be designed to pro-mote interdisciplinary understanding of natural and built environments by combining science, arts, and humanities. Use of music and dance should enhance both learning and subsequent involvement in conser-vation activities. Listening to music elevates endorphin levels, creating pleasurable feelings and stimulating learning. More research is needed to determine the types of active and passive art experiences that most effectively evoke changes in people's behavior in relation to the environ-ment, but we already know enough to sense the great potential of such activities for enhancing the effectiveness of environmental education.[17]

Environment, Design, and Restorative Emotions

During the last twenty years, an evolutionary perspective on human emotions has proven increasingly useful in psychiatry, medicine, and nutrition.[18] Deviating from the landscapes and diets of our ancestors often leads to undesirable outcomes. Patients recovering from surgery in hospitals with either views of natural vegetation or simulated views that depict natural scenes with water, recover more rapidly and are less anxious while recovering than patients with no access to natural views or who see only abstract designs.[19] If people spend even a few hours in forests, parks, and other places with trees, they experience increased immune function. The cause appears to be phytoncides, airborne chemicals that plants emit to protect themselves from insects and other herbivores and prevent rotting.[20]

If aesthetic responses evolved because they helped people solve life's problems, exposure to high quality environments should be restorative; it should reduce feelings of tension and stress. It does! More than one hundred studies show that stress reduction consistently emerges as a benefit of recreation in wild places.[21] Spending time in urban parks and other landscaped seminatural settings also reduces stress.[22] The savanna hypothesis predicts that feelings of relaxation and peacefulness should be especially strongly evoked by settings resembling savannas. University students often seek natural environments or urban settings dominated by natural elements (wooded urban parks, places offering views of natural landscapes, locations at edges of water) when they feel stressed or depressed. Today, however, stressed students are increasingly turning to social media to relax and connect with friends. The long-term consequences of this substitution are uncertain, but, as I have demonstrated, we have neural circuits honed to respond positively to specific landscape features; virtual environments should not be as effective as natural ones. We should experimentally test the strength and nature of our responses to real and virtual worlds before we assume that access to real worlds does not matter.

Designing Environmental Policies

Since the 1970s, the US government has developed ways to measure how people feel about their environment. Policy makers want to know how citizens feel about the effects that projects, like building dams or widening highways, will have on the surrounding environment. Using surveys and interviews, agencies try to assess the value people place on wilderness and unspoiled places, and their willingness to pay to create, preserve, and maintain them.

A good example is the way that concepts from evolutionary psychology are being used to help manage aesthetic resources on public land. In 1993 the US Forest Service published Agricultural Handbook Number 701, "Landscape Aesthetics: A Handbook for Scenery Management."[23] Its purpose is to assist land managers in implementing a "Scenery Management System" to improve visitors' physiological well-being as "an important byproduct of viewing *interesting and pleasant natural landscapes of high depth views and natural-appearing open spaces*" (emphasis in the original). The recommendations draw on psychological and physiological studies of people under stress, recovering in hospitals, and in recreational and other settings, that demonstrate the restorative properties of natural settings. The results of studies by Dimberg, Ulrich, and Simons, showing that heart rate (beats per minute) decreases when people view spatially open landscapes but not when they view spatially restricted environments are particularly important. Based on these results, the handbook recommends assessing the scenic attractiveness of landscapes and offers suggestions for enhancing the visibility of a landscape by altering plants and trees or buildings that obscure the view.

The Human-Animal Bond

Humans depend on animals, especially mammals, for food, fiber, and countless other commodities. We are attracted to scenes with large mammals. The big five (elephant, rhinoceros, cape buffalo, lion, and leopard) are the main attractions on African safaris. As we have seen, we have neural programs that are sensitive to animals and their movements. Stress is reduced when people observe arrays of moving animals. As expected, keeping pets is very pleasurable. The use of animals as companions may have begun before animals were domesticated.[24] Psychiatrists have long believed that animals can help minds and bodies heal, but until recently we had only anecdotal evidence. Today, many species, from dogs and rabbits to horses and llamas, are used as therapy animals both with children with autism and adults with chronic problems such as headaches, impaired vision, and dementia. Often expressionless and silent patients smile, laugh, and talk to the animals as well as the therapy animal's handlers. Speech, attention, and nonverbal expression of emotions are strikingly improved if people have a pet or regular access to a companion animal.[25] Why should this be so?

Our ancestors' success in hunting and gathering depended on gaining information about where animals were in the environment and communicating that information to other people. They would have benefited from trying to think like the animals they were hunting. That would have helped them understand and, hence, predict their behavior. Asking questions like "If I were a rabbit, where would I hide from predators?" would have been helpful. The success of such anthropomorphism may have favored our tendency to assume that animals really have conscious strategies.[26] Over the millennia, that mental habit would, in turn, have favored the development of totemism, which is a formal recognition of our kinship with animals. Bonding emotionally with animals was certain to evolve.

The Healing Garden

Three thousand years ago, in what is now Iraq, people tried using horticulture as emotional therapy. Persians created gardens designed to please multiple senses by combining beauty, fragrance, the sights and sounds of water, birdsong, and shade. In 1812, Benjamin Rush, a physician and professor of the Institute of Medicine and Clinical Practice at the University of Pennsylvania, noted that digging in a garden was one of the activities that distinguished male patients who recovered from mania from those that did not.[27] Rush inspired many public and private psychiatric hospitals to include gardening and farming activities in their programs. In the 1940s, the US Veterans Administration established hospitals to care for wounded servicemen and -women. Members of garden clubs and the horticultural industry brought flowers to the hospitals and introduced plant-based activities.[28] In the 1950s, Alice Burlingame, a trained psychiatrist, established horticulture therapy programs with volunteers from the National Farm and Garden Bureau; she also coauthored the first book on horticultural therapy.

In 1973, a group of horticultural therapy professionals established the Council for Therapy and Rehabilitation through Horticulture. Now called the American Horticultural Therapy Association, it has nearly one thousand members; most of them are registered professionals. A rich literature that describes the cognitive, psychological, social, and physical benefits of horticultural therapy supports their belief that working with plants promotes emotional, mental, and physical health and well-being. The organization publishes the *Journal of Therapeutic Horticulture* and supports a variety of activities.

We Are Regressing Backward

Although people have an inherent love of nature—what Edward O. Wilson calls *biophilia*—these feelings, like many other evolved traits, fail to flourish unless nurtured in early childhood. We can be introduced to nature via three avenues—direct, indirect, and vicarious. Direct experience is by far the most important for children's cognitive development and in their feelings about nature because it engages a wider range of problem-solving responses than the other two modes.[29] Yet, in the United States today the majority of most children's interactions with nature are via Twitter, YouTube, and other electronic media—*videophilia*. Such sedentary and solitary forms of interacting with nature negatively affect cognitive development, contribute to childhood obesity, and increase feelings of loneliness and depression.[30] We need to get young people out into nature. The campaign by Richard Louv, a Berkeley, California journalist, "No Child Left Inside," is pointing the way.[31] Louv has identified a serious mental problem, "nature deficit disorder," that afflicts a large proportion of children today.

Mayan children only four or five years old are familiar with more than one hundred species of plants; suburban children in the United States can name only a few.[32] Most undergraduate students at Northwestern University, in Evanston, Illinois, a suburb of Chicago, had heard of some of the local trees (birch, cedar, chestnut, fig, hickory, maple, oak, pine, and spruce), but fewer than half had any familiarity with alder, buckeye, linden, mountain ash, sweet gum, and tulip tree, all of which are common in the Evanston area. Lack of detailed, firsthand knowledge about particular species leads to loss of the meanings attached to them. We call that process backward regression.

The serious loss of knowledge about and, hence, appreciation of, nature that accompanies growing up in a modern technological society bodes ill for our efforts to preserve Earth's biodiversity. Indirect experiences of nature do not generate the emotional involvement with living organisms that appears to be necessary to motivate people to care about biodiversity and devote time and money to help preserve it. People cannot mourn the loss of something they did not know existed. We cannot expect people who have never experienced nature to appreciate it and be willing to work to save it.

Our Pleistocene Minds in Today's World

Our savanna-adapted ecological minds generally function quite well in modern industrial, urbanized societies. Our attraction to savanna vegetation and trees that are similar to those that dominate high-quality African savannas poses no problems for us today. The parks and gardens we design and the art we create that mimic our ancestral homeland enrich our lives. Our love of music has stimulated some of the finest expressions of human creativity. Making and listening to music brings great joy to the lives of people on all human cultures. We may sometimes become obsessed with our collections of things, but, by and large, collecting and categorizing things is a source of pleasure, not of problems.

Unfortunately, some of our savanna-derived biases do pose problems for us today. Our love of food has generated the rich and varied cuisines that so delight and sustain us, but our craving for sweets has allowed the food industry to make fortunes while we become increasingly obese and face an epidemic of type 2 diabetes. We can't control what is a normal, deeply rooted urge for a nutritious resource that was scarce during most of human history.

Our autonomic nervous system circuits evolved to respond strongly to stimuli that were relevant in the past (snakes, dangerous herbivores, and large carnivorous mammals) not to stimuli that pose greater threats today (firearms, electrical wires, speeding cars). We should be afraid of guns, driving fast, driving without a seat belt, and hair dryers near bathtubs, not of snakes and spiders. Efforts by public officials to strike fear in citizens using statistics and shocking photographs usually fail. Parents scold to deter their children from playing with matches or chasing a ball into the street, but when Chicago children were asked what they were most afraid of, they cited lions, tigers, and snakes. We find it very difficult to control those fears. We cannot eliminate them by consciously reminding ourselves that, for example, most snakes are actually harmless.

Our failure to fear truly dangerous objects and situations in today's environment causes serious social problems. We drive dangerously and kill one another at high rates on highways because we fail to recognize how dangerous driving a motor vehicle at high speeds on crowded roads really is. Weapons of all kinds, from handguns to nuclear bombs, rarely evoke a level of fear that matches their real hazard. The thrill of the hunt, combined with the power of modern firearms, easily leads to decimation of wildlife and high rates of homicide.

The consequences of failing to address the stark mismatch between the things we fear the most and the things that are most likely to kill us are serious. To deal with such problems as high accident rates and an epidemic of obesity and its associated type 2 diabetes, we must first recognize that they are biological problems. This tells us that solutions will not come easily because the urges that drive them are strong. We cannot rely on research to come up with silver bullets that will cause the problems to shrink or vanish. Enhanced public education to help us make better decisions will certainly help but is likely to yield only modest results. Alas, our increasing obesity is accompanied by more and more books on how to diet.

Clearly, the incentives we have chosen to employ are insufficient for the task. Speed limits lower accident rates but psychologically we do not fear driving fast. We slow down because we fear being arrested, not because we fear having an accident. We happily drive faster than the speed limit when we think we can get away with it. In addition to regulations, which are an essential part of reigning in our savanna minds, we need to devise and employ incentives that either directly affect our emotions or prevent us from yielding to them.

To combat obesity, for example, we could enact laws to protect children. We already require parents to enroll their children in school, have them immunized, and make them wear seat belts. We don't let them buy alcohol or cigarettes. Type 2 diabetes used to be called "adult onset," but it is increasing rapidly today in children along with their increasing obesity. Why not prevent them from having access to unhealthy doses of sugary drinks at school? We could ban unhealthy food in schools and require schools to provide daily physical education, which unfortunately, has been dropped in many school districts. We could require that mass-marketed junk food bear prominent health warning labels. Reducing the negative effects of such deeply rooted biological problems will be difficult, but we know that the consequences of failing to tackle them will become increasingly serious. Ghosts of the Pleistocene in our psyches are both good and bad.

Environmental Aesthetics and Progress

We understand more about how the world works than ever before, and our knowledge base grows exponentially. Unfortunately, this accumulation may not make us more sensible. As Edward O. Wilson suggested, we are drowning in information but starved for wisdom. Applying results of environmental aesthetics may improve human well-being, but ethical and political progress may lag behind. History tells us that knowledge mostly increases power; it has little influence on rationality or morality; our needs and emotions remain basically the same. The major decisions made by our ancestors during the Pleistocene are the same ones we fret over today. Yet, the improved understanding of ourselves that evolutionary psychology is giving us can be a source of inspiration.

An evolutionary perspective offers a rich and rewarding view of what it means to be a human being on Earth. Following Charles Darwin's lead we have developed a new perspective of life on Earth, its origins, and how the millions of species we share our small planet with came into being. Evolutionary biology has also given us answers to the deep questions people everywhere have asked. Who are we? Where did we come from? Where are we going? We now employ evolutionary perspectives to think creatively about the challenges our savanna ancestors faced, how they evolved to deal with them, and why their responses have left important legacies in our modern minds. We understand why we have a sense of beauty and awe, and why emotions drive our decisions.

We have much more to learn, but we can be certain that an evolutionary perspective will continue to yield new and interesting insights into our interactions with the complex environment we live in. Most reassuring is the good news that most of the things we like to do are good for us to engage in. The evolutionary mirror has told us why that is so. Only a few animals—elephants and apes—recognize themselves in a mirror. Humans have self-reflected for a long time, but we have an increasingly deeper understanding of the nature of the savanna ape whose reflection we view if we take time to look.

If we look carefully in a mirror, we can see a Boran hunter whistling to a honeyguide, a spreading acacia tree casting welcome midday shade, and, out of the corner of our eyes, a python hiding in the grass. We can recognize both our current selves and the savanna environment that, with its challenges and opportunities, molded us. It is a comforting vision. We can and should take pleasure in contemplating this inspiring glimpse of ourselves as part of the unfolding of life on Earth over nearly four billion years.

Acknowledgments

During the many years I have thought about and studied the evolutionary roots of human emotional responses to components of the environment, I have been fortunate to benefit from insights and advice from Judi Heerwagen. We worked together on environmental aesthetics for more than twenty years, and we have coauthored several papers on the subject. Her influence permeates this book. The first article I published on environmental aesthetics in 1980 was the result of being invited by Joan Lockard to contribute a chapter in her book on human evolution. Writing that article stimulated my desire to continue to explore the evolutionary roots of our responses to the environment. Lionel Tiger, then a program officer with the Harry Frank Guggenheim Foundation, urged me to continue my explorations of this theme; the foundation provided a small grant that supported our initial research in Kenya. My wife, Betty, was my constant companion during that research. She held the ends of tape measures, recorded data, took pictures of the trees, and kept a watchful eye out for lions and snakes, which were difficult to see even in short grass. She performed similar functions during our investigations of Japanese and European formal gardens, where predators were absent.

I have been fortunate to benefit from insights and advice of many colleagues. Useful comments on individual chapters have been provided by John Alcock, David Barash, George Brengelman, Gardner Brown, Eliot Brenowitz, Joseph Carroll, Sue Christian, Richard Coss, Emily Doolittle, John Edwards, Arne Öhman, Eric Pianka, Paul Rozin, William Searcy, Tore Slagsvold, Eric Alden Smith, and Robert Sommer. Ann Downer-Hazell, my talented editor, skillfully helped me liven my prose without putting too many words into my mouth. I owe special thanks to Leda Cosmides, Eric Dinerstein, and Marianne Kogon—they read the entire manuscript and offered a wealth of insightful comments. As is always the case, they bear no responsibility for residual errors or my inappropriate conclusions and interpretations.

Notes

Chapter One

1. Most scientists doubted that honeyguides really guided people to bees' nests until experiments by Isack and Reyer clearly demonstrated that they do (H. A. Isack and H. U. Reyer, "Honeyguides and Honey Gatherers: Interspecific Communication in a Symbiotic Relationship," *Science* 243 [1989]: 1343–46).
2. I describe my youthful experiences and my subsequent scientific career in G. H. Orians, "My Life with Birds," in *Leaders in Animal Behavior: The Second Generation*, ed. Lee Drickamer and Donald Dewsbury (Cambridge: Cambridge University Press, 2010), 405–27.
3. E. O. Wilson, *Sociobiology: The New Synthesis* (Cambridge, MA: Belknap Press, 1975). Wilson's book summarized the literature and stimulated decades of research in behavioral ecology.
4. R. Dawkins, *Unweaving the Rainbow* (London: Penguin, 1998).
5. E. Fehr and J. Russell, "Concepts of Emotions Viewed from a Prototype Perspective," *Journal of Experimental Psychology: General* 113 (1984): 464–86. The quote appears on page 464.
6. G. H. Orians, "Habitat Selection: General Theory and Applications to Human Behavior," in *Evolution of Human Social Behavior*, ed. J. S. Lockard (New York: Elsevier, 1980), 49–77. Being asked to write this article and the responses it generated stimulated me to pursue the research that eventually led to this book.

Chapter Two

1. A. G. Baumgarten, *Aesthetica* (Frankfurt: I. C. Kleyb, 1750); F. J. Kovach, *Philosophy of Beauty* (Norman: University of Oklahoma Press, 1974).
2. Charles Darwin, *The Expression of the Emotions in Man and Animals* (London: John Murray, 1872); W. James, "What Is an Emotion?," *Mind* 9 (1884): 188–205; W. Wundt, *Outlines of Psychology*, trans. C. H. Judd (Leipzig: Englemann, 1897).
3. T. Reid, *Essays on the Intellectual Powers of Man* (Edinburgh: Maclachian and Stewart, 1785).
4. For a good summary, see J. T. Cacioppo, L. G. Tassinary, and A. J. Fridlund, "The Skeletomotor System," in *Principles of Psychology: Physical, Social, and Inferential Elements*, ed. J. T. Cacioppo and L. G. Tassinary (Cambridge: Cambridge University Press, 1990).
5. For an excellent summary, see L. Cosmides and J. Tooby, "Evolutionary Psychology

and the Emotions," in *Handbook of Emotions*, ed. M. Lewis and J. M. Haviland-Jones, 2nd ed. (New York: Guilford Press, 2000), 91–115.

6. J. Appleton, *The Experience of Landscape* (New York: Wiley, 1975).

7. J. A. Byers, *American Pronghorn: Social Adaptations and the Ghosts of Predators Past* (Chicago: University of Chicago Press, 1997).

8. R. G. Coss and E. P. Charles, "The Role of Evolutionary Hypotheses in Psychological Research: Instincts, Affordances, and Relic Sex Differences," *Ecological Psychology* 16 (2004): 199–236; R. G. Coss and R. O. Goldthwaite, "The Persistence of Old Designs for Perception," in *Perspectives in Ethology*, ed. N. S. Thompson, vol. 11, *Behavioral Design* (New York: Plenum Press, 1995), 83–148; R. L. Sussman, J. T. Stern Jr., and W. L. Jungers, "Arboreality and Bipedality in the Hadar Hominids," *Folia Primatologica* 43 (1984): 113–56.

9. R. G. Coss and R. O. Goldthwaite, "The Persistence of Old Designs for Perception," in *Perspectives in Ethology*, ed. N. S. Thompson, vol. 11, *Behavioral Design* (New York: Plenum Press, 1995), 83–148

10. A good summary is provided by J. Tooby and L. Cosmides, "The Psychological Foundations of Culture," in *The Adapted Mind: Evolutionary Psychology and the Generation of Culture*, ed. J. Barkow, L. Cosmides, and J. Tooby (New York: Oxford University Press, 1992), 19–136.

11. J. H. Heerwagen and G. H. Orians, "The Ecological World of Children," in *Children and Nature: Psychological, Sociocultural, and Evolutionary Investigations*, ed. P. H. Kahn Jr. and S. R. Kellert (Cambridge, MA: MIT Press, 2002), 29–63; G. H. Orians and J. H. Heerwagen, "Evolved Responses to Landscapes," in *The Adapted Mind*, ed. J. H. Barkow, L. Cosmides, and J. Tooby (New York: Oxford University Press, 1992), 555–79.

12. D. E. Brown, *Human Universals* (New York: McGraw-Hill, 1991); R. B. Lee and I. Devore, eds., *Kalahari Hunter-Gatherers: Studies of the !Kung San and Their Neighbors* (Cambridge, MA: Harvard University Press, 1976); K. R. Hill et al., "Coresidence Patterns in Hunter-Gatherer Societies Show Unique Human Social Structure," *Science* 331 (2011): 1286–89; P. Shipman and A. Walker, "The Costs of Becoming a Predator," *Journal of Human Evolution* 18 (1989): 373–92.

13. M. Konner, "Hunter-Gatherer Infancy and Childhood: The !Kung and Others," in *Hunter-Gatherer Childhoods: Evolutionary, Developmental and Cultural Perspectives*, ed. B. S. Hewlett and M. E. Lamb (New Brunswick, NJ: Aldine Transaction, 2005), 19–64.

14. Ibid.

15. K. Hill and A. M. Hurtado, *Ache Life History: The Ecology and Demography of a Foraging People* (New York: Aldine de Gruyter, 1996).

16. M. E. P. Seligman, "On the Generality of the Laws of Learning," *Psychological Review* 77 (1970): 406–18.

17. J. Rydell, H. Roninen, and K. W. Philip, "Persistence of Bat Defense Reactions in High Arctic Moths (Lepidoptera)," *Proceedings of the Royal Society of London, Series B* 267 (2000): 553–57; R. G. Coss and R. O. Goldthwaite, "The Persistence of Old Designs for Perception," in *Perspectives in Ethology*, ed. N. S. Thompson, vol. 11, *Behavioral Design* (New York: Plenum Press, 1995), 83–148.

18. C. J. Raxworthy, "Reptiles," in *The Natural History of Madagascar*, ed. S. M. Goodman and J. P. Benstead (Chicago: University of Chicago Press, 2003), 934–51.

19. R. S. Ulrich, "Biophilia, Biophobia, and Natural Landscapes," in *The Biophilia Hypothesis*, ed. S. R. Kellert and E. O. Wilson (Washington, DC: Island Press, 1993), 73–137; E. O. Wilson, *Biophilia* (Cambridge, MA: Harvard University Press, 1984).

20. J. Diamond, "New Guineans and Their Natural World," in *The Biophilia Hypothesis*, ed. S. R. Kellert and E. O. Wilson (Washington, DC: Island Press, 1993), 25–171.

21. G. P. Nabhan, *Why Some Like It Hot: Food, Genes, and Cultural Diversity* (Washington, DC: Island Press, 2004).

Chapter Three

1. G. C. Williams, *Adaptation and Natural Selection: A Critique of Some Current Evolutionary Thought* (Princeton, NJ: Princeton University Press, 1966).

2. T. J. Wills et al., "Development of the Hippocampal Cognitive Map in Preweanling Rats," *Science* 328 (2010): 1573–76.

3. S. A. Lee, V. A. Sovrano, and E. S. Spelke, "Navigation as a Source of Geometric Knowledge: Young Children's Use of Length, Angle, Distance, and Direction in a Reorientation Task," *Cognition* 123 (2012): 144–61.

4. W. T. Powers, *Behavior: The Control of Perception* (Chicago: Aldine de Gruyter, 1973).

5. J. Tooby and L. Cosmides, "The Psychological Foundations of Culture," in *The Adapted Mind: Evolutionary Psychology and the Generation of Culture*, ed. J. Barkow, L. Cosmides, and J. Tooby (New York: Oxford University Press, 1992), 19–136.

6. A thorough synthesis of the literature on childhood and its evolution is M. Konner's *The Evolution of Childhood: Relationships, Emotion, Mind* (Cambridge, MA: Belknap Press, 2010); J. H. Heerwagen and G. H. Orians, "The Ecological World of Children," in *Children and Nature: Psychological, Sociocultural, and Evolutionary Investigations*, ed. P. H. Kahn Jr. and S. R. Kellert (Cambridge, MA: MIT Press, 2002), 29–63.

7. J. Appleton, *The Symbolism of Habitat* (Seattle: University of Washington Press, 1990).

8. J. J. Gibson, *The Ecological Approach to Visual Perception* (Boston: Houghton-Mifflin, 1979).

9. M. Kyttä, M. Kaaja, and L. Horelli, "An Internet-Based Design Game as a Mediator of Children's Environmental Visions," *Environment and Behavior* 36 (2004): 127–51.

10. H. R. Pulliam and C. Dunford, *Programmed to Learn: An Essay on the Evolution of Culture* (New York: Columbia University Press, 1980); M. E. P. Seligman, "On the Generality of the Laws of Learning," *Psychological Review* 77 (1970): 406–18.

11. D. E. Brown, *Human Universals* (New York: McGraw-Hill, 1991).

12. J. Von Uexkull, *Theoretical Biology* (New York: Harcourt, Brace, 1926).

13. J. Bowlby, *Attachment and Loss* (New York: Basic Books, 1980); M. Konner, *The Evolution of Childhood: Relationships, Emotion, Mind* (Cambridge, MA: Belknap Press, 2010).

Chapter Four

1. C. Finlayson et al., "The *Homo* Habitat Niche: Using the Avian Fossil Record to Depict Ecological Characteristics of Palaeolithic Eurasian Hominins," *Quaternary Science Review* 30 (2011): 1–8.

2. Captain R. B. Marcy, Report on Exploration and Survey of Route from Fort Smith,

Arkansas, to Santa Fe, New Mexico, Made in 1849, House ex. Doc. 45: 31st Cong., 1st sess., Public Document 577 (1849), quoted on page 89, in J. C. Malin, *The Grassland of North America: Prolegomena to Its History, with Addenda* (Lawrence, KS: James Malin, 1956).

3. M. L. Cody, ed., *Habitat Selection in Birds* (New York: Academic Press, 1985).

4. J. E. R. Staddon, "Learning as Inference," in *Evolution and Learning*, ed. R. C. Bolles and M. D. Beecher (Hillsdale, NJ: Erlbaum, 1988), 59–77 ; J. Tooby and L. Cosmides, "The Past Explains the Present: Emotional Adaptations and the Structure of Ancient Environments," *Ethology and Sociobiology* 11 (1990): 375–424.

5. G. H. Orians and J. H. Heerwagen, "Evolved Responses to Landscapes," in *The Adapted Mind*, ed. J. H. Barkow, L. Cosmides, and J. Tooby (New York: Oxford University Press, 1992), 555–79.

6. R. B. Zajonc, "Feeling and Thinking: Preferences Need No Inferences," *American Psychologist* 35 (1980): 151–75; R. M. Baron, "Social Knowing from an Ecological-Event Perspective: A Consideration of the Relative Domains of Power for Cognitive and Perceptual Modes of Knowing," in *Cognition, Social Behavior, and the Environment*, ed. J. H. Harvey (Hillsdale, NJ: Erlbaum, 1981), 61–89.

7. R. Kaplan and S. Kaplan, *The Experience of Nature: A Psychological Perspective* (New York: Cambridge University Press, 1989); R. S. Ulrich, "Visual Landscapes and Psychological Well-Being," *Landscape Research* 4 (1979): 17–23; R. S. Ulrich, "Aesthetic and Affective Response to Natural Environments," in *Human Behavior and Environment*, ed. I. Altman and J. F. Wohlwill, vol. 6, *Behavior and the Natural Environment* (New York: Plenum Press, 1983).

8. N. K. Humphrey, "Natural Aesthetics," in *Architecture for People*, ed. B. Mikellides (London: Studis Vista, 1980), 59–73; S. Kaplan, "Aesthetics, Affect, and Cognition: Environmental Preferences from an Evolutionary Perspective," *Environment and Behavior* 19 (1987): 3–32; S. Kaplan, "Environmental Preference in a Knowledge-Seeking, Knowledge-Using Organism," in *The Adapted Mind*, ed. J. H. Barkow, L. Cosmides, and J. Tooby (New York: Oxford University Press, 1992), 591–98; D. E. Berlyne, *Aesthetics and Psychobiology* (New York: Appleton-Century-Crofts, 1971); S. Kaplan and R. Kaplan, *Cognition and Environment: Functioning in an Uncertain World* (New York: Praeger, 1982).

9. J. Appleton, *The Experience of Landscape* (New York: Wiley, 1975).

10. K. Z. Lorenz, *King Solomon's Ring* (New York: Crowell, 1952). The quote is on page 181 of the 1964 edition.

11. S. Kaplan, "Environmental Preference in a Knowledge-Seeking, Knowledge-Using Organism," in *The Adapted Mind*, ed. J. H. Barkow, L. Cosmides, and J. Tooby (New York: Oxford University Press, 1992), 591–98.

12. K. Lynch, *The Image of the City* (Cambridge, MA: MIT Press, 1960).

13. S. Kaplan, "Environmental Preference in a Knowledge-Seeking, Knowledge-Using Organism," in *The Adapted Mind*, ed. J. H. Barkow, L. Cosmides, and J. Tooby (New York: Oxford University Press, 1992), 591–98.

14. J. Haidt, *The Happiness Hypothesis* (New York: Basic Books, 2006). The quote is on page 21.

Chapter Five

1. J. Bennett-Levy and T. Marteau, "Fear of Animals: What Is Prepared?," *British Journal of Psychology* 75 (1984): 37–42.
2. W. S. Agras, D. Sylvester, and D. Oliveau, "The Epidemiology of Common Fears and Phobias," *Comprehensive Psychiatry* 10 (1969): 151–56.
3. A. Öhman, "Face the Beast and Fear the Face: Animal and Social Fears as Prototypes for Evolutionary Analyses of Emotion," *Psychophysiology* 23 (1986): 123–45.
4. M. Davis and Y. Lee, "Fear and Anxiety: Possible Roles of the Amygdala and Bed Nucleus of the Striata Terminalis," *Cognition and Emotion* 12 (1998): 277–305; P. J. Lang, M. Davis, and A. Öhman, "Fear and Anxiety: Animal Models and Human Cognitive Psychophysiology," *Journal of Affective Disorders* 61 (2001): 137–59; S. Mineka, "The Role of Fear in Theories of Avoidance Learning, Flooding and Extinction," *Psychological Bulletin* 86 (1979): 985–1001; A. Öhman and S. Mineka, "The Malicious Serpent: Snakes as a Prototypical Stimulus for an Evolved Module of Fear," *Current Directions in Psychological Science* 12 (2003): 5–9; D. Viviani et al., "Oxytocin Selectivity Gates Fear Responses through Distinct Outputs from the Central Amygdala." *Science* 333 (2011): 104–7.
5. J. E. LeDoux, "Emotion Circuits in the Brain," *Annual Review of Neuroscience* 23 (2000): 155–84; A. Öhman and S. Mineka, "Fears, Phobias, and Preparedness: Toward an Evolved Module of Fear and Fear Learning," *Psychological Science* 108 (2001): 438–522; P. Rozin and E. B. Royzman, "Negativity Bias, Negativity Dominance, and Contagion," *Personality and Social Psychology Review* 5 (2001): 296–320.
6. L. A. Isbell, *The Fruit, the Tree, and the Serpent: Why We See so Well* (Cambridge, MA: Harvard University Press, 2009).
7. R. G. Coss, "The Role of Evolved Perceptual Biases in Art and Design," in *Evolutionary Aesthetics*, ed. E. Voland and K. Gammer (Berlin: Springer-Verlag, 2003), 69–130.
8. S. A. Kastner, P. De Weerd, and L. G. Ungerleider, "Texture Segregation in the Human Visual Cortex: A Functional MRI Study," *Journal of Neurophysiology* 83 (2000): 2453–57.
9. T. Okusa, R. Kakgi, and N. Osaka, "Cortical Activity Related to Cue-Invariant Shape Perception in Humans," *Neuroscience* 98 (2000): 615–24.
10. A. Öhman, A. Flykt, and F. Esteves, "Emotion Drives Attention: Detecting the Snake in the Grass," *Journal of Experimental Psychology: General* 130 (2001): 466–78.
11. L. A. Isbell, *The Fruit, The Tree, and The Serpent: Why We See so Well* (Cambridge, MA: Harvard University Press, 2009).
12. H. Miller, "The Cobra, India's 'Good Snake,'" *National Geographic* 138 (1970): 393–408.
13. J. Diaz-Bolio, *The Geometry of the Maya and Their Rattlesnake Art* (Mérida: Area Maya-Mayan Area, 1987).
14. L. A. Isbell, *The Fruit, The Tree, and The Serpent: Why We See So Well* (Cambridge, MA: Harvard University Press, 2009).
15. J. White, "Bites and Stings from Venomous Animals: A Global Overview," *Therapeutic Drug Monitor* 22 (2000): 65–68.
16. M. Cook and S. Mineka, "Selective Associations in the Observational Conditioning of

Fear in Rhesus Monkeys," *Journal of Experimental Psychology: Animal Behavior Processes* 16 (1990): 372–89.

17. V. LoBue and J. S. DeLoache, "Detecting the Snake in the Grass: Attention to Fear-Relevant Stimuli by Adults and Young Children," *Psychological Science* 19 (2008): 284–89; J. S. DeLoache and V. LoBue, "The Narrow Fellow in the Grass: Human Infants Associate Snakes and Fear," *Developmental Science* 12 (2009): 201–7.

18. A. J. Tomarken, S. K. Sutton, and S. Mineka, "Fear-Relevant Illusory Correlations: What Types of Associations Promote Judgmental Bias?," *Journal of Abnormal Psychology* 104 (1995): 312–26.

19. R. G. Coss, "The Role of Evolved Perceptual Biases in Art And Design," in *Evolutionary Aesthetics*, ed. E. Voland and K. Gammer (Berlin: Springer-Verlag, 2003), 69–130.

20. C. E. Osgood, G. J. Suci, and P. H. Tannenbaum, *The Measurement of Meaning* (Urbana: University of Illinois Press, 1957).

21. C. L. Larson, J. Aronoff, and J. J. Stearns, "The Shape of Threat: Simple Geometric Forms Evoke Rapid and Sustained Capture of Attention," *Emotion* 7 (2007): 526–53; C. L. Larson et al., "Recognizing Threat: A Simple Geometric Shape Activates Neural Circuitry for Threat Detection." *Journal of Cognitive Neuroscience* 21 (2009): 1523–35.

22. A. Treves and L. Naughton-Treves, "Risk and Opportunity for Humans Coexisting with Large Carnivores," *Journal of Human Evolution* 36 (1999): 275–82.

23. R. G. Coss, "The Role of Evolved Perceptual Biases in Art And Design," in *Evolutionary Aesthetics*, ed. E. Voland and K. Gammer (Berlin: Springer-Verlag, 2003), 69–130.

24. J. Mellaart, *Çatal Hüyük: A Neolithic Town in Anatolia* (London: Thames and Hudson, 1968).

25. A. Wolfe and B. Sleeper, *Wild Cats of the World* (New York: Crown, 1995).

26. R. Desimone, "Face-Selective Cells in the Temporal Cortex of Monkeys," *Journal of Cognitive Neuroscience* 3 (1991): 1–8; C. Bruce, R. Desimone, and C. G. Gross, "Visual Properties of Neurons in a Polysensory Area in Superior Temporal Sulcus of the Macaque," *Journal of Neurophysiology* 46 (1981): 369–84; K. Tanaka, "Representation of Visual Features of Objects in the Inferotemporal Cortex," *Neural Networks* 9 (1996): 1459–75; D. I. Perrett, E. T. Rolls, and W. Caan, "Visual Neurons Responsiveness to Faces in the Monkey Temporal Cortex," *Experimental Brain Research* 47 (1982): 329–42.

27. R. G. Coss, U. Ramakrishnan, and J. Schank, "Recognition of Partially Concealed Leopards by Wild Bonnet Macaques (*Macaca radiata*): The Role of the Spotted Coat," *Behavioural Processes* 68 (2005): 145–63.

28. R. G. Coss, "The Perceptual Aspects of Eye-Spot Patterns and Their Relevance to Gaze Behaviour," in *Behaviour Studies in Psychiatry*, ed. C. Hutt and S. J. Hutt (London: Pergammon Press, 1970), 121–24.

29. R. G. Coss, *Mood Provoking Visual Stimuli: Their Origins and Applications* (Los Angeles: University of California Press, 1965); R. G. Coss, "The Perceptual Aspects of Eye-Spot Patterns and Their Relevance to Gaze Behaviour," in *Behaviour Studies in Psychiatry*, ed. C. Hutt and S. J. Hutt (London: Pergammon Press, 1970), 121–24; J. Topal and V. Csányi, "The Effect of Eye-Like Schema on Shuttling Activities of Wild House Mice (*Mus musculus domesticus*): Context-Dependent Threatening Aspects of the Eye-spot Patterns," *Animal Learning and Behavior* 22 (1994): 96–102.

30. R. G. Coss, "The Role of Evolved Perceptual Biases in Art and Design," in *Evolutionary Aesthetics*, ed. E. Voland and K. Gammer (Berlin: Springer-Verlag, 2003), 69–130.

31. A. T. Jersild and F. B. Holmes, *Children's Fears* (New York: Bureau of Publications, Teachers College, Columbia University, 1935); T. G. R. Bower, *Development in Infancy* (San Francisco: Freeman, 1974); P. K. Smith, "The Ontogeny of Fear in Children," in *Fear in Animals and Man*, ed. W. Sluckin (New York: Van Nostrand Reinhold, 1979).

32. L. Berk, *Child Development*, 4th ed. (Boston: Allyn and Bacon, 1997).

33. A. J. Fridlund, *Human Facial Expressions: An Evolutionary View* (San Diego: Academic Press, 1994).

34. C. Darwin, *The Expression of the Emotions in Animals and Man* (London: John Murray, 1872).

35. R. A. Hinde, *Biological Bases of Human Social Behavior* (New York: McGraw-Hill, 1974); W. K. Redican, "An Evolutionary Perspective on Human Facial Displays," in *Emotion in the Human Face*, ed. P. Eckman (New York: Cambridge University Press, 1982), 212–80; R. J. Andrews, "Evolution and Facial Expression," *Science* 142 (1963): 1034–41.

36. S. Hrdy, *The Woman Who Never Evolved* (Cambridge, MA: Harvard University Press, 1981); E. K. Roberts et al., "A Bruce Effect in Wild Geladas," *Science* 335 (2012): 1222–25.

37. I. M. Marks, *Fears, Phobias and Rituals: Panic, Anxiety and Their Disorders* (New York: Oxford University Press, 1987).

38. E. J. Mazurski et al., "Conditioning with Facial Expressions of Emotions: Effects of CS Sex and Age," *Psychophysiology* 33 (1996): 416–25.

39. G. A. Morgan and H. N. Ricciutti, "Infants' Responses to Strangers during the First Year," in *Determinants of Infant Behavior*, ed. B. M. Foss (London: Methun, 1967); P. K. Smith, "The Ontogeny of Fear in Children," in *Fear in Animals and Man*, ed. W. Sluckin (New York: Van Nostrand Reinhold, 1979); M. Daly and M. Wilson, "Evolutionary Social Psychology and Family Homicide," *Science* 242 (1988): 519–24.

40. T. M. Horner, "Two Methods of Studying Danger Reactivity in Infants: A Review," in *Annual Progress in Child Psychiatry and Child Development*, ed. S. Chess and A. Thomas (New York: Bruner/Mazel, 1981), 78–98.

41. G. W. Bronson, "Infants' Reactions to Unfamiliar Persona and Novel Objects," *Monographs of the Society for Research in Child Development* 37, no. 3 (1972).

42. S. Zegans and L. S. Zegans, "Fear of Strangers in Children and the Orienting Reaction," *Behavioral Science* 17 (1972): 407–19.

43. A. Maurer, "What Children Fear," *Journal of Genetic Psychology* 106 (1965): 265–77.

44. M. G. Haselton and D. M. Buss, "Error Management Theory: A New Perspective on Biases in Cross-Sex Mind Reading," *Journal of Personality and Social Psychology* 78 (2000): 81–91.

45. R. E. Jackson, "Falling Towards a Theory of the Vertical-Horizontal Illusion," in *Studies in Perception and Action*, ed. H. Left and K. L. Marsh (Hillsdale, NJ: Erlbaum, 2005), 8:241–44; R. E. Jackson and L. K. Cormack, "Evolved Navigation Theory and the Environmental Vertical Illusion," *Evolution and Human Behavior* 29 (2008): 299–304.

46. S. Scarr and P. Salapetek, "Patterns of Fear Development during Infancy," *Merrill-Palmer Quarterly* 16 (1970): 53–90.

47. N. J. King, D. I. Hamilton, and T. H. Ollendick, *Children's Phobias: A Behavioral Perspective* (New York: Wiley, 1988); I. M. Marks, *Fears, Phobias and Rituals: Panic, Anxiety*

and Their Disorders (New York: Oxford University Press, 1987); I. M. Marks and M. G. Gelder. "Different Ages of Onset in Varieties of Phobia," *American Journal of Psychiatry* 123 (1966): 218–21.

48. D. Lieberman, *The Evolution of the Human Head* (Cambridge, MA: Belknap Press, 2011). This book provides a thorough review of the development of the human head and the unfolding of its neural circuits.

49. J. W. Anderson, "Attachment Behaviors Out of Doors," in *Ethological Studies of Child Behavior*, ed. N. Blurton-Jones (Cambridge: Cambridge University Press, 1972).

50. J. Bowlby, *Attachment and Loss* (New York: Basic Books, 1980); C. Garvey, *Children's Play* (Cambridge, MA: Harvard University Press, 1990); B. L. White et al. "Competence and Experience," in *The Structure of Experience*, ed. I. C. Uzgiris and F. Weizmann (New York: Plenum Press, 1977), 15–152; R. A. Chase, "Toys and Infant Development: Biological, Psychological, and Social Factors," *Children's Environments* 9 (1992): 3–12.

51. S. Scarr and P. Salapetek, "Patterns of Fear Development during Infancy," *Merrill-Palmer Quarterly* 16 (1970): 53–90: P. K. Smith, "The Ontogeny of Fear in Children," in *Fear in Animals and Man*, ed. W. Sluckin (New York: Van Nostrand Reinhold, 1979), 164–68.

52. P. Muris, H. Merckelbach, and R. Collaris, "Common Childhood Fears and Their Origins," *Behavioral Research and Therapy* 35 (1997): 929–37.

53. R. Lapouse and M. A. Monk, "Fears and Worries in a Representative Sample of Children," *American Journal of Orthopsychiatry* 29 (1959): 803–18.

54. W. S. Agras, D. Sylvester, and D. Oliveau. "The Epidemiology of Common Fears and Phobias," *Comprehensive Psychiatry* 10 (1969): 151–56.

55. P. Muris, H. Merckelbach, and R. Collaris, "Common Childhood Fears and Their Origins," *Behavioral Research and Therapy* 35 (1997): 929–93; T. H. Ollendick, J. L. Matson, and W. J. Heisel, "Fears in Children and Adolescents: Normative Data," *Behavior Research and Therapy* 23 (1985): 465–67; H. Angelino, J. Dollins, and E. V. Mech, "Trends in the 'Fears and Worries' of School Children as Related to Socio-Economic Status and Age," *Journal of Genetic Psychology* 89 (1956): 263–76.

56. E. R. Hagman, "A Study of Fears of Children of Pre-School Age," *Journal of Experimental Education* 1 (1932): 110–30; F. B. Holmes, "An Experimental Study of Fears of Young Children," *Child Development Monographs* 20 (1935): 167–296.

57. N. J. King, D. I. Hamilton, and T. H. Ollendick, *Children's Phobias: A Behavioral Perspective* (New York: Wiley, 1988); H. Angelino, J. Dollins, and E. V. Mech, "Trends in the 'Fears and Worries' of School Children as Related to Socio-Economic Status and Age," *Journal of Genetic Psychology* 89 (1956): 263–76; J. W. Croake, "Fears of Children," *Human Development* 12 (1969): 239–74; T. H. Ollendick, J. L. Matson, and W. J. Heisel, "Fears in Children and Adolescents: Normative Data," *Behavior Research and Therapy* 23 (1985): 465–67; *Diagnostic and Statistical Manual of Mental Disorders: DSM-IV*, 4th ed. (Washington, DC: American Psychiatric Association, 1994).

58. A. Öhman, U. Dimberg, and L.-G. Öst, "Animal and Social Phobias: Biological Constraints on Learned Fear Responses," in *Theoretical Issues in Behavior Therapy*, ed. S. Reiss and R. R. Bootzin (New York: Academic Press, 1985), 123–78.

59. M. G. Haselton and D. Nettle, "The Paranoid Optimist: An Integrative Evolutionary

Model of Cognitive Biases," *Personality and Social Psychology Review* 10, no. 1 (2006): 47–66.

60. D. Kahneman, E. Diener, and N. Schwarz, eds., *Well-Being: The Foundations of Hedonic Psychology* (New York: Russell Sage Foundation, 1999); D. Kahneman, and A. Tversky, eds., *Choices, Values and Frames* (New York: Cambridge University Press, 2000); D. Kahneman, "A Perspective on Judgment and Choice: Mapping Bounded Rationality," *American Psychologist* 58, no. 9 (2003): 697–720; J. L. Knetsch, "Asymmetric Valuation of Gains and Losses and Preference Order Assumptions," *Economic Inquiry* 33 (1995): 134–41; P. Rozin and E. B. Royama, Negativity Bias, Negativity Dominance, and Contagion," *Personality and Social Psychology Review* 5 (2001): 296–320.

61. R. J. McNally, "Preparedness and Phobias: A Review," *Psychological Bulletin* 101 (1987): 283–303; K. Hugdahl and A. C. Kärker, "Biological vs. Experiential Factors in Fear-Conditioning," *Behavioral Research and Therapy* 19 (1981): 109–15.

62. K. Hugdahl, "Electrodermal Conditioning to Potentially Phobic Stimuli: Effects of Instructed Extinction," *Behavioral Research and Therapy* 16 (1978): 315–21; S. Hygge and A. Öhman, "Modeling Processes in the Acquisition of Fears: Vicarious Electrodermal Conditioning to Fear-Relevant Stimuli," *Journal of Personality and Social Psychology* 36 (1978): 271–79.

63. M. Cook and S. Mineka, "Selective Associations in the Observational Conditioning of Fear in Rhesus Monkeys," *Journal of Experimental Psychology: Animal Behavior Processes* 16 (1990): 372–89.

64. A. Öhman and J. J. F. Soares, "On the Automatic Nature of Phobic Fear: Conditioned Electrodermal Responses to Masked Fear-Relevant Stimuli," *Journal of Abnormal Psychology* 103 (1993): 231–40; A. Öhman, "Face the Beast and Fear the Face: Animal and Social Fears as Prototypes for Evolutionary Analyses of Emotion," *Psychophysiology* 23 (1986): 123–45.

65. P. K. Smith, "The Ontogeny of Fear in Children," in *Fear in Animals and Man*, ed. W. Sluckin (New York: Van Nostrand Reinhold, 1979), 164–68.

Chapter Six

1. V. Komar and A. Melamid, *Painting by the Numbers: Komar and Melamid's Scientific Guide to Art*, ed. J. Wypijewski (New York: Farrar, Straus and Giroux, 1997).

2. E. Dissanayake, "Komar and Melamid Discover Pleistocene Taste," *Philosophy and Literature* 22 (1998): 486–96.

3. N. K. Humphrey, "Natural Aesthetics," in *Architecture for People*, ed. B. Mikellides (London: Studis Vista, 1980), 59–73; S. Kaplan, "Aesthetics, Affect, and Cognition: Environmental Preferences from an Evolutionary Perspective," *Environment and Behavior* 19 (1987): 3–32; S. Kaplan, "Environmental Preference in a Knowledge-Seeking, Knowledge-Using Organism," in *The Adapted Mind*, ed. J. H. Barkow, L. Cosmides, and J. Tooby (New York: Oxford University Press, 1992), 591–98; S. Kaplan and R. Kaplan, *Cognition and Environment: Functioning in an Uncertain World* (New York: Praeger, 1982); D. E. Berlyne, *Aesthetics and Psychobiology* (New York: Appleton-Century-Crofts, 1971).

4. A. T. Grove and O. Rackham, *The Nature of Mediterranean Europe: An Ecological History* (New Haven, CT: Yale University Press, 2001), 193.

5. Sir F. Crisp, *Medieval Gardens*, ed. C. C. Patterson (London: Bodley Head, 1924).

6. P. L. Carpenter, T. D. Walker, and F. O. Lamphear, *Plants in the Landscape* (San Francisco: W. H. Freeman, 1975).

7. Yi-Fu Tuan, *Topophilia: A Study of Environmental Perception, Attitudes, and Values* (Englewood Cliffs, NJ: Prentice-Hall, 1974).

8. Lady M. Shikibu, *The Tale of Genji*, trans. Arthur Waley (London: Allen and Unwin, 1935).

9. D. M. Van Gelderen and J. R. P. van Hoey Smith, *Conifers* (Portland, OR: Timber Press, 1986).

10. J. D. Vertrees, *Japanese Maples* (Forest Grove, OR: Timber Press, 1978).

11. J. H. Heerwagen and G. H. Orians, "Humans, Habitats, and Aesthetics," in *The Biophilia Hypothesis*, ed. S. R. Kellert and E. O. Wilson (Washington, DC, Island Press, 1993), 138–72.

12. P. Thiel, "Why Dig Japan Today?" (Vancouver: Program Exhibition Notes, Roots of Japanese Architecture, Fine Arts Gallery, University of British Columbia, November 16–December 3, 1966).

13. L. B. Alberti, *De Re Aedificatoria* (Florence, 1485); L. B. Alberti, *Ten Books on Architecture*, trans. J. Leoni (1724).

14. H. Repton, *The Art of Landscape Gardening* (Boston: Houghton- Mifflin, 1907), 96; G. H. Orians and J. H. Heerwagen, "Evolved Responses to Landscapes," in *The Adapted Mind*, ed. J. H. Barkow, L. Cosmides, and J. Tooby (New York: Oxford University Press, 1992), 555–79.

15. H. Repton, *The Art of Landscape Gardening* (Boston: Houghton-Mifflin, 1907), 97.

16. H. Repton, *Observations on the Theory and Practice of Landscape Gardening* (London: Taylor, 1803); H. Repton, *The Art of Landscape Gardening* (Boston: Houghton-Mifflin, 1907).

17. Lady E. C. S.-Wortley, *Travels in the United States, etc., during 1849 and 1850* (London: Richard Bentley, 1851).

18. A. Danto, "Can It Be the 'Most Wanted' Even if Nobody Wants It?," in *Painting by the Numbers: Komar and Melamid's Scientific Guide to Art*, ed. J. Wypijewski (New York: Farrar, Straus, and Giroux, 1997).

19. R. Sommer and J. Summit, "Cross-Rankings of Tree Shape," *Ecological Psychology* 8 (1996): 327–41.

20. R. Sommer, "Further Cross-National Studies of Tree Form Preference," *Ecological Psychology* 9 (1997): 153–60.

21. V. I. Lohr and C. H. Pearson-Mims, "Responses to Scenes with Spreading, Rounded, and Conical Tree Forms," *Environment and Behaviour* 38 (2006): 667–68.

22. R. G. Coss and M. Moore, "Precocious Knowledge of Trees as Antipredator Refuge in Preschool Children: An Examination of Aesthetics, Attributive Judgments, and Relic Sexual Dinichism," *Ecological Psychology* 14: (2002), 181–222.

23. A. E. Bigelow, "Hiding in Blind and Sighted Children," *Development and Psychopathology* 3 (1991): 301–10; A. Bridges and J. Rowles, "Young Children's Projective Abilities: What Can a Monster See?" *Educational Psychology* 5 (1985): 251–66.

24. U. Ramakrishnan and R. G. Coss, "A Comparison of the Sleeping Behavior of Three Sympatric Primates: A Preliminary Report," *Folia Primatologica* 72 (2001a): 51–53; U. Ramakrishnan and R. G. Coss, "Strategies Used by Bonnet Macaques (*Macaca radiata*) to Reduce Predation While Sleeping," *Primates* 432 (2001b): 193–206.

25. N. Altman, *Sacred Trees* (San Francisco: Sierra Club Books, 1993).

26. P. Fuller, "The Geography of Mother Nature," in *The Iconography of Landscape,* ed. D. Cosgrove and S. Daniels (Cambridge: Cambridge University Press, 1988), 11–31.

27. B. P. Meier and M. D. Robinson, "Why the Sunny Side Is Up: Associations between Affect and Vertical Position," *Psychological Science* 15 (2004): 243–47.

28. E. D. Benson, J. L. Hansen, and A. L. Schwartz, "Water View and Residential Property Values," *Appraisal Journal* 68 (2000): 260–71; E. D. Benson et al., "Pricing Residential Amenities: The Value of a View," *Journal of Real Estate Finance and Economics* 16 (1998): 55–73.

29. R. H. Plattner and T. J. Campbell, "A Study of the Effect of Water View on Site Value," *Appraisal Journal* 46 (1978): 20–28.

30. J. Luttik, "The Value of Trees, Water, and Open Space as Reflected by House Prices in The Netherlands," *Landscape and Urban Planning* 48 (2000): 161–67.

31. E. L. David and W. B. Lord, "Determinants of Property Value on Artificial Lakes," unpublished report (University of Wisconsin, College of Agriculture, Madison, 1969).

32. Yi-Fu Tuan, *Topophilia: A Study of Environmental Perception, Attitudes, and Values* (Englewood Cliffs, NJ: Prentice-Hall, 1974).

Chapter Seven

1. Athenaeus (ca. 200 AD), *The Deipnosophists,* trans. C. B. Glick, 7 vols. (Cambridge, MA: Loeb Classical Library, 1961); D. P. Patterson, "Plant Exploration," *Proceedings of the International Plant Propagators Society* 20 (1970): 251–57.

2. L. Kass, *The Hungry Soul* (New York: Free Press, 1994).

3. P. Quinnett, *Darwin's Bass* (Kansas City, MO: Andrews McMeel, 1998).

4. R. W. Young, "Evolution of the Human Hand: The Role of Throwing and Clubbing," *Journal of Anatomy* 202 (2003): 165–74; William Calvin explores this in detail in *The Throwing Madonna: Essays on the Brain* (Boston: McGraw-Hill, 1983).

5. W. S. Laughlin, "Hunting: An Integrating Biobehavior System and Its Evolutionary Importance," in *Man the Hunter,* ed. R. B. Lee and I. Devore (Chicago: Aldine, 1968).

6. M. Gurven, H. Kaplan, and M. Guttierrez, "How Long Does It Take to Become a Proficient Hunter? Implication on the Evolution of Delayed Growth," *Journal of Human Evolution* 51 (2006): 454–70.

7. K. Hawkes, "Why Do Men Hunt? Benefits for Risky Choices," in *Risk and Uncertainty in Tribal and Peasant Economies,* ed. E. Cashdan (Boulder, CO: Westview Press, 1990), 145–66.

8. M. G. Bicchieri, ed., *Hunters and Gatherers Today* (New York: Holt, Rinehart and Winston, 1972).

9. M. J. Sharps et al., "Memory for Animal Tracks: A Possible Cognitive Artifact of Human Evolution," *Journal of Psychology* 136 (2002): 469–92.

10. A. Le Hardÿ de Beaulieu, *An Illustrated Guide to Maples* (Portland, OR: Timber Press, 2001).

11. A. Wulf , *The Brother Gardeners: Botany, Empire and the Birth of an Obsession* (New York: Knopf, 2008).

12. J. Haviland-Jones et al., "An Environmental Approach to Positive Emotion: Flowers," *Evolutionary Psychology* 3 (2005): 104–32.

13. G. P. Nabhan, *Why Some Like It Hot: Food, Genes, and Cultural Diversity* (Washington, DC: Island Press, 2004).

14. P. Rozin, "Why We Eat What We Eat, and Why We Worry about It," *Bulletin of the American Academy of Arts and Sciences L* (1997): 26–48.

15. C. R. Peters, E. M. O'Brien, and E. O. Box, "Plant Types and Seasonality of Wild-Plant Foods, Tanzania to Southwestern Africa: Resources for Models of the Natural Environment," *Journal of Human Evolution* 13 (1984): 397–414.

16. T. Johns, *With Bitter Herbs They Shall Eat It: Chemical Ecology and the Origins of Human Diet and Medicine* (Tucson: University of Arizona Press, 1990).

17. G. H. Orians and N. P. Pearson, "On the Theory of Central Place Foraging," in *Analysis of Ecological Systems*, ed. D. V. Horn, R. D. Mitchell, and G. R. Stairs (Columbus: Ohio State University Press, 1979) 155–77.

18. J. M. Broughton, "Prehistoric Human Impacts on California Birds: Evidence from the Emeryville Shellmound Avifauna," *Ornithological Monographs* no. 26 (2004); American Ornithologists' Union.

19. P. Rozin, "Why We Eat What We Eat, and Why We Worry about It," *Bulletin of the American Academy of Arts and Sciences L* (1997): 26–48.

20. I. C. Uzgiris, "Ordinality in the Development of Schemas for Relating to Objects," in *Exceptional Infant-Normal Infant*, ed. J. Helmuth (New York: Bruner/Mazel, 1967), 317–34.

21. R. G. Coss, S. Ruff, and T. Simms, "All That Glistens: II. The Effects of Reflective Surface Finishes on the Mouthing Activity of Infants and Toddlers," *Ecological Psychology* 15 (2003): 197–21; I. C. Uzgiris, "Ordinality in the Development of Schemas for Relating to Objects," in *Exceptional Infant-Normal Infant*, ed. J. Helmuth (New York: Bruner/Mazel, 1967), 317–34.

22. J. V. Neel, "Lessons from a 'Primitive' People," *Science* 170 (1970): 815–22.

23. P. Rozin, "Getting to Like the Burn of Chili Pepper: Biological, Psychological and Cultural Perspectives," in *Chemical Senses: Irritation*, ed. B. G. Green, J. R. Mason, and M. R. Kare, vol. 2 (New York: Marcel Dekker, 1990), 231–69; A. A. Blake, "Flavor Perception and the Learning of Food Preferences," in *Flavour Perception*, ed. A. J. Taylor and D. D. Roberts (Oxford: Blackwell, 2004), 172–98.

24. P. Rozin, "Why We Eat What We Eat, and Why We Worry about It," *Bulletin of the American Academy of Arts and Sciences L* (1997): 26–48.

25. C. Darwin, *The Expression of the Emotions in Man and Animals* (London: John Murray, 1872). Page citations are from the 1998 edition annotated by Paul Ekman.

26. C. Nemeroff and P. Rozin, "You Are What You Eat: Applying the Demand-Free 'Impressions' Unique to an Unacknowledged Belief Ethos," *Journal of Psychological Anthropology* 17 (1989): 50–69.

27. P. Rozin, "Why We Eat What We Eat, and Why We Worry about It," *Bulletin of the American Academy of Arts and Sciences L* (1997): 26–48.

28. P. Rozin, J. Haidt, and C. R. McCauley, "Disgust," in *Handbook of Emotions*, ed. M. Lewis and J. M. Haviland-Jones, 2nd ed. (New York: Gilford Press, 2000), 637–53; J. Haidt, *The Happiness Hypothesis* (New York: Basic Books, 2006); J. Haidt and C. Joseph, "Intuitive Ethics: How Innately Prepared Intuitions Generate Culturally Variable Virtues," *Daedalus* 133 (2004): 55–66.

29. J. Haidt, *The Happiness Hypothesis* (New York: Basic Books, 2006).

30. H. A. Chapman et al., "In Bad Taste: Evidence for the Oral Origins of Moral Disgust," *Science* 323 (2009): 1222–26.

31. P. Rozin, J. Haidt, and K. Fincher, "From Oral to Moral," *Science* 323 (2009): 1179–80.

32. K. Thomas, *Man and the Natural World: A History of the Modern Sensibility* (New York: Pantheon Books, 1983).

33. P. W. Sherman and J. Billings, "Darwinian Gastronomy: Why We Use Spices," *BioScience* 49 (1998): 453–63.

34. G. P. Nabhan, *Why Some Like It Hot: Food, Genes, and Cultural Diversity* (Washington, DC: Island Press, 2004); P. W. Sherman and G. A. Hash, "Why Vegetable Recipes Are Not Very Spicy," *Evolution and Human Behavior* 22 (2001): 147–63.

35. M. Eisenstein, "Of Beans and Genes," *Nature* 468 (2010): 513–15.

36. S. H. Katz, "Fava Bean Consumption: A Case for the Co-Evolution of Genes and Culture," in *Food and Evolution*, ed. M. Harris and E. B. Ross (Philadelphia: Temple University Press, 1987), 133–59; M. Kurlansky, *Salt: A World History* (New York: Penguin Books, 2003).

37. R. Wrangham and T. Nishida, "*Aspilia* spp. Leaves: A Puzzle in the Feeding Behavior of Wild Chimpanzees," *Primates* 24 (1983): 276–82.

38. V. Wobber, B. Hare, and R. Wrangham, "Great Apes Prefer Cooked Food," *Journal of Human Evolution* 55 (2008): 340–48.

39. R. Wrangham, *Catching Fire: How Cooking Made Us Human* (New York: Basic Books, 2009).

40. M. Profet, "The Evolution of Pregnancy Sickness as Protection to the Embryo against Pleistocene Teratogens," *Evolutionary Theory* 8 (1988): 177–90; M. Profet, "Pregnancy Sickness as Adaptation: A Deterrent to Maternal Ingestion of Teratogens," in *The Adapted Mind*, ed. J. Barkow, L. Cosmides, and J. Tooby (Oxford: Oxford University Press, 1992), 327–66.

41. S. M. Flaxman and P. W. Sherman, "Morning Sickness: A Mechanism for Protecting Mother and Embryo," *Quarterly Review of Biology* 75 (2000): 113–48.

42. C. Stanford, *Chimpanzee and Red Colobus: The Ecology of Predator and Prey* (Cambridge, MA: Harvard University Press, 1998).

43. J. H. Barkow, *Darwin, Sex, and Status: Biological Approaches to Mind and Culture* (Toronto: University of Toronto Press, 1989).

44. R. Hames, "Sharing among the Yanomamö: Part 1: The Ethics of Risk," in *Risk and Uncertainty in Tribal and Peasant Economies*, ed. E. Cashdan (Boulder: University of Colorado Press, 1990), 89–106.

45. H. Kaplan and H. Hill, "Food Sharing among Ache Foragers: Tests of Explanatory Hypotheses," *Current Anthropology* 26 (1985): 223–39.

46. Ibid.

47. P. D. Dwyer and M. Minnegal, "Hunting in Lowland Tropical Rainforest: Towards a Model of Non-Agricultural Subsistence," *Human Ecology* 19 (1991): 187–212; P. M. Kaberry, "Political Organization in the Northern Abelam," *Politics in New Guinea*, ed. R. M. Berndt and P. Lawrence (Nedlands: University of Western Australia Press, 1971); A. Weiner, *The Trobrianders of Papua New Guinea* (New York: Holt, Rinehart and Winston, 1988).

48. M. Young, *Fighting with Food* (Cambridge: Cambridge University Press, 1971).

Chapter Eight

1. B. Krause, *The Great Animal Orchestra: Finding the Origins of Music in the World's Wild Places* (New York: Little, Brown, 2012).

2. Ibid.

3. P. Gouk, *Music Healing in Cultural Contexts* (Aldershot: Ashgate, 2000).

4. J. M. Standley, "The Effect of Contingent Music to Increase Non-Nutritive Sucking of Premature Infants," *Pediatric Nursing* 26 (2000): 493–95, 498–99; J. M. Standley, "The Effect of Music-Reinforced Non-Nutritive Sucking on Feeding Rate of Premature Infants," *Journal of Pediatric Nursing* 18 (2003): 169–73.

5. A. Lomax, *Folk Song Style and Culture* (Piscataway, NJ: Transaction Publishers, 1978).

6. C. Darwin, *The Descent of Man, and Selection in Relation to Sex* (New York: D. Appleton, 1872), quotes on 877, 878, and 880.

7. G. Miller, "Evolution of Human Music through Sexual Selection," in *The Origins of Music*, ed. N. L. Wallin, B. Merker, and S. Brown (Cambridge, MA: MIT Press, 2000), 329–60.

8. C. Levi-Strauss, *L'Homme nu* (Paris: Plon, 1971).

9. S. Mithen, *The Singing Neanderthals: The Origins of Music, Language, Mind and Body* (London: Weidenfeld and Nicolson, 2005).

10. G. Miller, "Evolution of Human Music through Sexual Selection," in *The Origins of Music*, ed. N. L. Wallin, B. Merker, and S. Brown (Cambridge, MA: MIT Press, 2000), 329–60.

11. D. Kunej and I. Turk, "New Perspectives on the Beginnings of Music: Archaeological and Musicological Analysis of a Middle Paleolithic Bone 'Flute,'" in *The Origins of Music*, ed. N. L. Wallin, B. Merker, and S. Brown (Cambridge, MA: MIT Press, 2000), 235–68.

12. N. J. Conard, M. Malina, and S. C. Münzel, "New Flutes Document the Earliest Musical Tradition in Southwestern Germany," *Nature* 460 (2009): 737–40; F. D'Errico et al., "Archaeological Evidence for the Emergence of Language, Symbolism, and Music—An Alternative Inter-Disciplinary Perspective," *Journal of World Prehistory* 17 (2003): 1–70.

13. J. Tooby and L. Cosmides, "Evolutionary Psychology and the Emotions," in *Handbook of Emotions*, ed. M. Lewis and J. M. Haviland-Jones, 2nd ed. (New York: Guilford Press, 1990), 91–115.

14. J. Tooby and L. Cosmides, "The Psychological Foundations of Culture," in *The Adapted Mind: Evolutionary Psychology and the Generation of Culture*, ed. J. Barkow, L. Cosmides,

and J. Tooby (New York: Oxford University Press, 1992), 19–136; S. Pinker, *The Blank Slate: The Modern Denial of Human Nature* (New York: Penguin, 2002).

15. R. Dunbar, *Grooming, Gossip and the Evolution of Language* (Cambridge, MA: Harvard University Press, 1996).

16. Ibid.; J. Molino, "Toward an Evolutionary Theory of Music and Language," in *The Origins of Music*, ed. N. L. Wallin, B. Merker, and S. Brown (Cambridge, MA: MIT Press, 2000), 165–76.

17. W. L. Benzon, *Beethoven's Anvil: Music in Mind and Culture* (New York: Basic Books, 2001); B. Krause, *The Great Animal Orchestra: Finding the Origins of Music in the World's Wild Places* (New York: Little, Brown, 2012); V. Geist, *Life Strategies, Human Evolution, Environmental Design: Toward a Biological Theory of Health* (New York: Springer-Verlag, 1978).

18. V. Gazzola and C. Keysers, "The Observation and Execution of Actions Share Motor and Somatosensory Voxels in All Tested Subjects: Single-Subject Analyses of Unsmoothed fMRI Data," *Cereb Cortex* 19, no. 6 (2008): 1239–55, www.pubmedcentral.nih.gov/articlerender.fcgi?tool=pmcentrez&artid=2677653.

19. E. M. Thomas, *The Old Way: A Story of the First People* (New York: Farrar, Straus and Giroux, 2006).

20. E. Doolittle, "Crickets in the Concert Hall: A History of Animals in Western Music," *Transcultural Music Review* 12 (2008).

21. C. Sachs, *The Wellsprings of Music*, ed. Jaap Kunst (The Hague: Martin Nijhoff, 1962).

22. R. M. Seyfarth and D. L. Cheney, "Behavioral Mechanisms Underlying Vocal Communication in Nonhuman Primates," *Animal Learning and Behavior* 25 (1997): 249–67; D. L. Cheney and R. M. Seyfarth, *Baboon Metaphysics: The Evolution of a Social Mind* (Chicago: University of Chicago Press, 2007).

23. S. Blackmore, *The Meme Machine* (Oxford: Oxford University Press, 1999).

24. E. Dissanayke, *Art and Intimacy: How the Arts Began* (Seattle: University of Washington Press, 2000); E. Dissanayake, "Antecedents of the Temporal Arts in Early Mother-Infant Interaction," in *The Origins of Music*, ed. N. L. Wallin, B. Merker, and S. Brown (Cambridge, MA: MIT Press, 2000), 389–410.

25. E. Terhardt, "Gestalt Principles and Music Perception," in *Auditory Processing of Complex Sounds*, ed. W. A. Yost and C. S. Watson (Hillsdale, NJ: Erlbaum, 1987), 157–66; S. Trehub, "Human Processing Predispositions and Musical Universals," in *The Origins of Music*, ed. N. L. Wallin, B. Merker, and S. Brown (Cambridge, MA: MIT Press, 2000), 427–48; C. Trevarthen, "Musicality and the Intrinsic Motive Pulse: Evidence from Human Psychology and Infant Communication," special issue, *Musicae Scientiae* (1999–2000): 155–215.

26. T. Geissmann, "Gibbon Songs and Human Music from an Evolutionary Perspective," in *The Origins of Music*, ed. N. L. Wallin, B. Merker, and S. Brown (Cambridge, MA.: MIT Press, 2000), 103–23.

27. R. V. Alatalo, C. Glynn, and A. Lundberg, "Singing Rate and Female Attraction in the Pied Flycatcher: An Experiment," *Animal Behavior* 39 (1990): 601–3; B. Kempenaers, G. R. Verheyen, and A. A. Dhondt, "Extrapair Paternity in the Blue Tit (*Parus caeruleus*): Female Choice, Male Characteristics, and Offspring Quality," *Behavioral Ecology* 8 (1997): 481–92; C. K. Catchpole, "Sexual Selection and the Evolution of Complex

Songs among European Warblers of the Genus *Acrocephalus*," *Behaviour* 74 (1980): 149–66; C. K. Catchpole, J. Dittami, and B. Leisler, "Differential Responses to Male Song Repertoires in Female Songbirds Implanted with Oestradiol," *Nature* 312 (1984): 563–64; B. Catchpole, B. Leisler, and J. Dittami, "Sexual Differences in the Responses of Captive Great Reed Warblers (*Acrocephalus arundinaceus*) to Variations in Song Structure and Repertoire Size," *Ethology* 73 (1986): 69–77; C. K. Catchpole and P. J. B. Slater, *Bird Song: Biological Themes and Variations* (Cambridge: Cambridge University Press, 1995); C. K. Catchpole. and P. K. McGregor, "Sexual Selection, Song Complexity and Plumage Dimorphism in European Buntings of the Genus *Emberiza*," *Animal Behavior* 33 (1985): 1149–66; H. Lampe and G.-P. Saetre, "Female Pied Flycatchers Prefer Males with Larger Song Repertoires," *Proceedings of the Royal Society, London, B* 262 (1995): 163–67; W. A. Searcy, "Song Repertoire Size and Female Preferences in Song Sparrows," *Behavioral Ecology and Sociobiology* 14 (1984): 281–86; W. A. Searcy and K. Yasukawa, "Song and Female Choice," in *Ecology and Evolution of Acoustic Communication in Birds*, ed. D. I. Rubenstein and R. W. Wrangham (Ithaca, NY: Cornell University Press, 1996), 175–200; F. E. Wasserman and J. A. Cigliano, "Song Output and Stimulation of the Female in White-Throated Sparrows," *Behavioral Ecology and Sociobiology* 29 (1991): 55–59.

28. K. Payne, P. Tyack, and R. Payne, "Progressive Changes in the Songs of Humpback Whales (*Megaptera novaeangliae*): A Detailed Analysis of Two Seasons in Hawaii," in *Communication and Behavior of Whales*, ed. R. Payne (Boulder, CO: Westview Press, 1983), 9–57; K. Payne and R. Payne, "Large Scale Changes over 19 Years in Song of Humpback Whales in Bermuda," *Zeitschrift für Tierpsychologie* 68 (1985): 89–114; K. Payne, "The Progressively Changing Songs of Humpback Whales: A Window on the Creative Process in a Wild Animal," in *The Origins of Music*, ed. N. L. Wallin, B. Merker, and S. Brown (Cambridge, MA: MIT Press, 2000), 135–50.

29. S. Nowicki, S. Peters, and J. Podos, "Song Learning, Early Nutrition and Sexual Selection in Songbirds," *American Zoologist* 38 (1998): 179–90; S. Nowicki, W. A. Searcy, and S. Peters, "Brain Development, Song Learning and Mate Choice in Birds: A Review and Experimental Test of the 'Nutritional Stress Hypothesis,'" *Journal of Computational Physics A* 188 (2002): 1003–14.

30. W. A. Searcy and S. Nowicki, *The Evolution of Animal Communication: Reliability and Deception in Signaling Systems* (Princeton, NJ: Princeton University Press, 2005).

31. N. Neave et al., "Male Dance Moves That Catch a Woman's Eye," *Biology Letters* 7 (2011): 221–24.

32. G. Miller, "Evolution of Human Music through Sexual Selection," in *The Origins of Music*, ed. N. L. Wallin, B. Merker, and S. Brown (Cambridge, MA: MIT Press, 2000), 329–60.

33. N. Chomsky, *Aspects of the Theory of Syntax* (Cambridge, MA: MIT Press, 1965); N. Chomsky, *Rules and Representations* (Oxford: Blackwell, 1980).

34. D. Falk, "Hominid Brain Evolution and the Origin of Music," in *The Origins of Music*, ed. N. L. Wallin, B. Merker, and S. Brown (Cambridge, MA: MIT Press, 2000), 197–216.

35. T. Bever and R. Chiarello, "Cerebral Dominance in Musicians and Nonmusicians," *Science* 185 (1974): 537–39.

36. C. Sachs, *The Wellsprings of Music*, ed. Jaap Kunst (The Hague: Martin Nijhoff, 1962).

37. Ibid.

38. Ibid.

39. J. Sergent, E. Zuck, S. Terriah, and B. MacDonald, "Distributed Neural Network Underlying Musical Sight-Reading and Keyboard Performance," *Science* 257 (1992): 106–10; W. Chen et al., "Functional Mapping of Human Brain during Music Imagery Processing," *NeuroImage* 3 (1996): S205.

40. C. Sachs, *The Wellsprings of Music*, ed. Jaap Kunst (The Hague: Martin Nijhoff, 1962).

41. R. Jourdain, *Music, the Brain, and Ecstasy* (New York: Avon Books, 1998).

42. Ibid.

43. D. Huron, *Sweet Anticipation: Music and the Psychology of Expectation* (Cambridge, MA: MIT Press, 2006).

44. L. F. Baptista and R. A. Keister, "Why Bird Song Is Sometimes Like Music," *Perspectives in Biology and Medicine* 48 (2005): 426–43.

45. K. Payne, "The Progressively Changing Songs of Humpback Whales: A window into the Creative Process in a Wild Animal," in *The Origins of Music*, ed. N. L. Wallin, B. Merker, and S. Brown (Cambridge, MA: MIT Press, 2000), 135–50.

46. C. Hartshorne, *Born to Sing: An Interpretation and World Survey of Bird Song* (Bloomington: Indiana University Press, 1973).

47. C. K. Catchpole and P. J. B. Slater, *Bird Song: Biological Themes and Variations* (Cambridge: Cambridge University Press, 1995).

Chapter Nine

1. L. Tiger, *The Pursuit of Pleasure* (New York: Little, Brown, 1992).

2. A. Gilbert, *What the Nose Knows: The Science of Scent in Everyday Life* (New York: Crown, 2008).

3. S. A. Goff and H. J. Klee, "Plant Volatile Compounds: Sensory Cues for Health and Nutritional Value?," *Science* 311 (2006): 815–91.

4. L. M. Nijssen et al., *Volatile Compounds in Food*, vol. 7, supp. 1 (Zeist: TNO Nutrition and Food Research Institute, 1997).

5. G. F. Oster and E. O. Wilson, *Caste and Ecology in the Social Insects* (Princeton, NJ: Princeton University Press, 1978).

6. D. H. Janzen, "Why Fruits Rot, Seeds Mold, and Meat Spoils," *American Naturalist* 111 (1977): 691–713.

7. C. Wedekind et al., "MHC-Dependent Mate Preferences in Humans," *Proceedings of the Royal Society: Biological Sciences* 266 (1995): 245–49.

8. R. J. Braidwood et al., "Symposium: Did Man Once Live by Beer Alone?" *American Anthropologist* 55 (1953): 515–26.

9. P. E. McGovern, *Ancient Wine: The Search for the Origin of Viniculture* (Princeton, NJ: Princeton University Press, 2006).

10. W. James, *The Varieties of Religious Experience* (London: Longmans, Green, 1902), 377.

11. R. Dudley, "Evolutionary Origins of Human Alcoholism in Primate Frugivory," *Quarterly Review of Biology* 75 (2000): 3–15.

12. Ibid.; R. Dudley, "Ethanol, Fruit Ripening, and the Historical Origins of Human Alcoholism and Primate Frugivory," *Integrative and Comparative Biology* 44 (200): 315–

32; D. Stephens and R. Dudley, "The Drunken Monkey Hypothesis," *Natural History* (December 2004/January 2005).

13. D. J. Levey, "The Evolutionary Ecology of Ethanol Production and Alcoholism," *Integrative and Comparative Biology* 44 (2004): 284–89.

14. D. M. Stoddard, "The Role of Olfaction in the Evolution of Human Sexuality: An Hypothesis," *Man* 21 (1986): 514–20; D. M. Stoddard, *The Scented Ape: The Biology and Culture of Human Odour* (Cambridge: Cambridge University Press, 1990).

15. G. H. Dodd, "The Molecular Dimension in Perfumery," in *Perfumery: The Psychology and Biology of Fragrance*, ed. S. Van Toller and G. H. Dodd (London: Chapmen and Hall, 1991), 19–46.

16. R. R. Calkin and J. S. Jellinek, *Perfumery: Practice and Principles* (New York: Wiley, 1994).

17. C. D. Daly and R. S. White, "Psychic Reactions to Olfactory Stimuli," *British Journal of Medical Psychology* 10 (1930): 70–87.

18. E. E. Slosson, "A Lecture Experiment in Hallucinogens," *Psychological Review* 6 (1899): 407–8.

19. A. Gilbert, *What the Nose Knows: The Science of Scent in Everyday Life* (New York: Crown, 2008).

20. Ibid., 25.

Chapter Ten

1. T. Johns, *With Bitter Herbs They Shall Eat: Chemical Ecology and the Origins of Human Diet and Medicine* (Tucson: University of Arizona Press, 1990).

2. S. H. Riesenberg, "Magic and Medicine in Ponape," *Southwest Journal of Anthropology* 4 (1948): 406–29.

3. T. Gladwin, *East Is a Big Bird: Navigation and Logic on Puluwat Atoll* (Cambridge, MA: Harvard University Press, 1970).

4. S. Kaplan, "Environmental Preference in a Knowledge-Seeking, Knowledge-Using Organism," in *The Adapted Mind*, ed. J. H. Barkow, L. Cosmides, and J. Tooby (New York: Oxford University Press, 1992), 591–98.

5. D. L. Medin, N. O. Ross, and D. G. Cox, *Culture and Resource Conflict: Why Meanings Matter* (New York: Russell Sage Foundation, 2006).

6. N. K. Humphrey, "Natural Aesthetics," in *Architecture for People*, ed. B. Mikellides (London: Studio Vista, 1980), 59–73. The quote is on page 64.

7. A. K. Thomason, *Luxury and Legitimation Royal Collecting in Ancient Mesopotamia* (Aldershot: Ashgate, 2005); A. Bounia, *The Nature of Classical Collecting. Collectors and Collections, 100 BCE—100 CE* (Aldershot: Ashgate, 2004).

8. D. F. Sherry, "Food Storage by Birds and Mammals," *Advances in the Study of Behavior* 15 (1985): 153–88; C. C. Smith and O. J. Reichman, "The Evolution of Food Caching by Birds and Mammals," *Annual Review of Ecology and Systematics* 15 (1984): 329–51.

9. I. P. Pavlov, "The Reflex of Purpose," in *Lectures on Conditioned Reflexes*, vol. 1. (London: Lawrence and Wishart, 1963).

10. S. W. Anderson, H. Damaio, and A. R. Damasio, "A Neural Basis for Collecting Behavior in Humans," *Brain* 128 (2005): 201–12.

11. A. Wulf, *Brother Gardeners: Botany, Empire and the Birth of an Obsession* (New York: Knopf, 2008).

12. C. Markham, *Colloquies on the Simples and Drugs of India by Garcia da Orta* (London: Henry Sotheran, 1913).

13. T. Birkhead, *A Brand-New Bird* (New York: Basic Books, 2003).

14. B. Krause, *The Great Animal Orchestra: Finding the Origins of Music in the World's Wild Places* (New York: Little, Brown, 2012).

Chapter Eleven

1. C. L. Apicella et al. "Social Networks and Cooperation in Hunter-Gatherers," *Nature* 481 (2012): 497–501.

2. A. W. Delton et al., "Evolution of Direct Reciprocity under Uncertainty Can Explain Human Generosity in One-Short Encounters," *Proceedings of the National Academy of Sciences* 108 (2011): 1335–40.

3. D. E. Brown, *Human Universals* (New York: McGraw-Hill, 1991).

4. Starlings benefit from additional information about the state of the environment when required to make sequential decisions about prey but not when they must choose between two simultaneously available prey items. E. Freiden and A. Kacelnik, "Rational Choice, Context Dependence, and the Value of Information in European Starlings (*Sturnus vulgaris*)," *Science* 334 (2011): 1000–1002.

5. D. Kahneman, *Thinking, Fast and Slow* (New York: Farrar, Straus and Giroux, 2011).

6. Yi-Fu Tuan, *Topophili: A Study of Environmental Perceptions, Attitudes, and Values* (Englewood Cliffs, NJ: Prentice-Hall, 1974).

7. C. Levi-Strauss, *La Pensée Sauvage* (Paris: Plon, 1962).

8. W. Burkert, *Creation of the Sacred: Tracks of Biology in Early Religion* (Cambridge, MA: Harvard University Press, 1998); S. Pinker, *The Stuff of Thought: Language as a Window into Human Nature* (New York: Penguin, 2007).

9. L. Cosmides and J. Tooby, *Evolutionary Psychology: Theoretical Foundations*, in *Encyclopedia of Cognitive Science* (London: Macmillan, 2003); S. Pinker, *The Blank Slate: The Modern Denial of Human Nature* (New York: Penguin, 2002).

10. J. G. Kingsolver and D. W. Pfennig, "Patterns and Power of Phenotypic Selection in Nature," *BioScience* 57 (2007): 561–72.

11. F. E. Kuo and A. F. Taylor, "A Potential Natural Treatment for Attention-Deficit/Hyperactivity Disorder: Evidence from a National Study," *American Journal of Public Health* 94 (2004): 1580–86.

12. Ibid.; A. F. Taylor and F. E. Kuo, "Children with Attention Deficits Concentrate Better after Walk in the Park," *Journal of Attention Disorders* 12 (2006): 402–9.

13. S. Kaplan and J. F. Talbot, "Psychological Benefits of a Wilderness Experience," in *Human Behavior and Environment*, ed. I. Altman and J. F. Wohlwill, vol. 6, *Behavior and the Natural Environment* (New York: Plenum Press, 1983).

14. T. Hartig, M. Mang, and G. W. Evans, "Restorative Effects of Natural Environmental Experiences," *Environment and Behavior* 23 (1991): 3–26; R. Hockey, ed., *Stress and Fatigue in Human Performance* (New York: Wiley, 1983); D. C. Glass and J. E. Singer, *Urban Stress: Experiments on Noise and Social Stressors* (New York: Academic Press, 1972).

15. R. S. Ulrich, "Biophilia, Biophobia, and Natural Landscapes," in *The Biophilia Hypothesis*, ed. S. R. Keller and E. O. Wilson (Washington, DC: Island Press, 1995), 73–137.

16. S. R. Kellert, J. H. Heerwagen, and M. L. Mader, eds., *Biophilic Design: The Theory, Science, and Practice of Bringing Buildings to Life* (New York: Wiley, 2008).

17. S. K. Jakobson, M. D. McDuff, and M. C. Monroe, *Conservation Education and Outreach Techniques* (New York: Oxford University Press, 2006).

18. G. C. Williams and R. M. Nesse, "The Dawn of Darwinian Medicine," *Quarterly Review of Biology* 66 (1991): 1–21; R. Nesse, "Evolutionary Explanations of Emotions," *Human Nature* 1 (1991): 261–89; R. Nesse and G. C. Williams, *Why We Get Sick: The New Science of Darwinian Medicine* (New York: Times Books, 1996).

19. R. S. Ulrich, "View through a Window May Influence Recovery from Surgery," *Science* 224 (1984): 420–21; R. S. Ulrich, O. Lundén, and J. L. Eltinge, "Effects of Exposure to Nature and Abstract Pictures on Patients Recovering from Heart Surgery," *Psychophysiology* 30, supp. 1 (1993): 7–26.

20. Q. Li et al., "A Forest Bathing Trip Increases Human Natural Killer Activity and Expression of Anti-Cancer Proteins in Female Subjects," *Journal of the Biological Regulation of Homeostatic Agents* 22 (2008): 45–55; Q. Li et al., "Visiting a Forest, but Not a City, Increases Human Natural Killer Activity and Expression of Anti-Cancer Proteins," *International Journal of Immunopathological Pharmacology* 21 (2008): 117–27.

21. R. C. Knopf, "Human Behavior, Cognition, and Affect in the Natural Environment," in *Handbook of Environmental Psychology*, ed. D. Stokols and I. Altman (New York: Wiley, 1987), 783–825; R. S. Ulrich, U. Dimberg, and B. L. Driver, "Psychophysiological Indicators of Leisure Benefits," in *Benefits of Leisure*, ed. B. L. Driver, P. J. Brown, and G. L. Peterson (State College, PA: Venture, 1991).

22. R. S. Ulrich and D. Addoms, "Psychological and Recreational Benefits of a Neighborhood Park," *Journal of Leisure Research* 13 (1981): 43–65; S. Kaplan, "The Restorative Benefits of Nature: Toward an Integrative Framework," *Journal of Environmental Psychology* 15 (1995): 169–82; S. Kaplan and J. F. Talbot, "Psychological Benefits of a Wilderness Experience," in *Human Behavior and Environment*, ed. I. Altman and J. F. Wohlwill, vol. 6, *Behavior and the Natural Environment* (New York: Plenum Press, 1983); H. W. Schroeder, "Environment, Behavior, and Design Research on Urban Forests," in *Advances in Environment, Behavior, and Design*, ed. E. H. Zube and G. T. Moore, vol. 2 (New York: Plenum Press, 1989).

23. C. Frances and C. Cooper-Marcus, "Places People Take Their Problems," in *Proceedings of the 22nd Annual Conference of the Environmental Design Research Association*, ed. J. Urbina-Soria, P. Ortega-Andeane, and R. Bechtel (Oklahoma City, OK: EDRA, 1991); US Department of Agriculture, Forest Service, *Landscape Aesthetics: A Handbook for Scenery Management* (Washington, DC: US Government Printing Office, 1993).

24. A. Katcher, "Man and the Living Environment: An Excursion into Cyclical Time," in *New Perspectives in Our Lives with Companion Animals*, ed. A. Katcher and A. Beck (Philadelphia: University of Pennsylvania Press, 1983).

25. A. Katcher and G. Wilkins, "Dialogue with Animals: Its Nature and Culture," in *The Biophilia Hypothesis*, ed. S. R. Kellert and E. O. Wilson (Washington, DC: Island Press, 1993), 173–97; A. Katche et al., "Looking, Talking, and Blood Pressure: The Physi-

ological Consequences of Interaction with the Living Environment," in *New Perspectives on Our Lives with Companion Animals,* ed. A. Katcher and A. Beck (Philadelphia: University of Pennsylvania Press, 1983); A. M. Beck, "Use of Animals in the Rehabilitation of Psychiatric Inpatients," *Psychological Reports* 58 (1986): 63–66; A. Katcher and G. Wilkins, "Dialogue with Animals: Its Nature and Culture," in *The Biophilia Hypothesis,* ed. S. R. Kellert and E. O. Wilson (Washington, DC Island Press, 1993), 173–97.

26. D. C. Dennett, *Kinds of Minds* (New York: Basic Books, 1996).

27. B. Rush, *Medical Inquiries and Observations upon Diseases of the Mind* (Philadelphia: Kimber and Richardson, 1812).

28. P. D. Reif, "Agriculture and Health Care: The Care of Plants and Animals for Therapy and Rehabilitation in the United States," in *Farming for Health,* ed. J. Hassink and M. van Dijk (Wageningen: Springer, 2006), 309–43.

29. S. R. Kellert, "Children's Affiliations with Nature: Structure, Development, and the Problem and Environmental Generational Amnesia," in *Children and Nature: Psychological, Sociocultural, and Evolutionary Investigations,* ed. P. H. Kahn Jr. and S. R. Kellert (Cambridge, MA: MIT Press, 2002), 93–116.

30. P. A. Zaradic and O. R. W. Pergams, "Videophilia: Implications for Childhood Development and Conservation," *Journal of Developmental Processes* 2 (2007): 130–44; O. R. W. Pergams and P. A. Zaradic, "Evidence for a Fundamental and Pervasive Shift Away from Nature-Based Recreation," *Proceedings of the National Academy of Sciences USA* 105 (2008): 2295–300.

31. R. Louv, *Last Child in the Woods: Saving Our Children from Nature-Deficit Disorder* (Chapel Hill, NC: Algonquin Books, 2005).

32. J. W. Dougherty, "Salience and Relativity in Classification," *American Ethnologist* 5 (1978): 665–80; B. Stross, "Acquisition of Botanical Terminology by Tzeltal Children," in *Meaning in Mayan Languages,* 107–41, ed. M. Edmonson (The Hague: Mourton, 1973), 1075–42.

Illustration Credits

1.1. International Bee Research Association, ibra.org. **1.2a.** Brian M. Wood, Yale University. **1.2b.** Joanna Eede. **1.3.** Special Collections Research Center, University of Chicago Library. **2.1.** Special Collections Research Center, University of Chicago Library. **2.2.** Tom and Pat Leeson, Ardea.com. **2.3.** © Michael Rothman 1997. **2.4.** Kalpesh Lathigra/World Wildlife Fund-UK **3.1.** Owen Humphries/PA/AP. **3.2.** Michael Amendolia. **3.3.** Image courtesy of Charles Nelson, PhD, Boston Children's Hospital; photo by Margaret Lampert. **4.1.** Reprinted from *Journal of Human Evolution*, 57/2, P. Utrilla et al., A paleolithic map from 13,660 calBP: engraved stone blocks from the Late Magdalenian in Abauntz Cave (Navarra, Spain), Fig. 7, Copyright 2009, with permission from Elsevier. **4.2.** Reprinted from *Journal of Human Evolution*, 57/2, P. Utrilla et al., A paleolithic map from 13,660 calBP: engraved stone blocks from the Late Magdalenian in Abauntz Cave (Navarra, Spain), Fig. 7, Copyright 2009, with permission from Elsevier. **4.3.** Martin Schoeller/AUGUST. **5.1.** AP Photo/Channi Anand. **5.2.** Composite image courtesy of Richard Coss. **5.3.** Bridgeman-Giraudon/Art Resource, NY. **5.4.** Photo courtesy of John Schoen. **6.1.** Courtesy of Ronald Feldman Fine Arts, New York/www.feldmangallery .com. **6.2.** bpk, Berlin/Staatliche Museen/Joerg P. Anders/Art Resource, NY. **6.3.** Gordon Orians. **6.4.** Gordon Orians. **6.5.** Gordon Orians. **6.6.** Courtesy of Arnold Arboretum Horticultural Library. © President and Fellows of Harvard College. Arnold Arboretum Archives. **6.7.** Gordon Orians. **6.8.** Gordon Orians. **6.9.** Jarrold Publishing/The Art Archive at Art Resource, NY. **6.10.** Gordon Orians. **6.11a.** Courtesy of the Smithsonian Institution Libraries, Washington, D.C. **6.11b.** Courtesy of the Smithsonian Institution Libraries, Washington, D.C. **6.12.** Gordon Orians **7.1.** Gianni Dagli Orti/The Art Archive at Art Resource, NY. **7.2.** James McDonald, 1922. Reproduction from a black-and-white glass lantern slide. Museum of New Zealand Te Papa Tongarewa. **7.3.** Reprinted from the *Victoria County History of Kent*, London, 1912. **7.4.** Paul Eckman, Ph.D./Paul Ekman Group, LLC. **7.5.** AP/Ho Chon In. **8.1.** Gordon Orians. **8.2.** Mark Newman/FLPA/Minden Pictures. **8.3.** Cornell Library of Ornithology. **8.4.** Library of Congress. **9.1.** Nick Gordon, naturepl.com. **9.2.** Bodleian Library, University of Oxford. **9.3.** Special Collections Research Center, University of Chicago Library **10.1.** AP/Toby Talbot. **10.2.** © Illustrated London News Ltd/Everett Collection. **10.3.** Hanford Declassification Project/ Department of Energy. **10.4.** Tommy Fabianus/Demotix. **10.5.** Tim Laman. **11.1.** Gordon Orians. **11.2.** AP/David Tipling/Solent News/Rex Features.

Index

Page numbers in italics refer to illustrations.